Negative Ecologies

The publisher and the University of California Press
Foundation gratefully acknowledge the generous support
of the Ralph and Shirley Shapiro Endowment Fund
in Environmental Studies.

Negative Ecologies

Fossil Fuels and the Discovery of the Environment

David Bond

UNIVERSITY OF CALIFORNIA PRESS

University of California Press
Oakland, California

Library of Congress Cataloging-in-Publication Data

Names: Bond, David, 1979- author.
Title: Negative ecologies : fossil fuels and the discovery of the environment /
    David Bond.
Description: Oakland, California : University of California Press, [2022] |
    Includes bibliographical references and index.
Identifiers: LCCN 2021046117 (print) | LCCN 2021046118 (ebook) |
    ISBN 9780520386778 (cloth) | ISBN 9780520386785 (paperback) |
    ISBN 9780520386792 (epub)
Subjects: LCSH: Fossil fuels—Environmental aspects—North America. |
    Ecology—Environmental aspects—North America. | Climatic changes—
    North America.
Classification: LCC TD887.F69 B66 2022 (print) | LCC TD887.F69 (ebook) |
    DDC 363.738/7097—dc23/eng/20211012
LC record available at https://lccn.loc.gov/2021046117
LC ebook record available at https://lccn.loc.gov/2021046118

31 30 29 28 27 26 25 24 23 22
10 9 8 7 6 5 4 3 2 1

To William and Meredith,
for the revolutions they bring

Contents

Illustrations

Introduction

The Promise and Predicament of Crude Oil

Crude oil is the world's greatest commodity. Whether by measure of revenue, trade, or even mass utility, for the past century crude oil has claimed an undisputed place atop the pedestal of first and foremost. Nothing else comes close. Every corner of the inhabited world bears the practical imprint of crude oil, as motorbikes, buses, ships, and planes now generalize an easy mobility of people and things. The professional buying and selling of oil, whether in barrels or in futures, outstrips every other market transaction, and private fortunes beyond the wealth of many nations now ride on the fluctuating price of oil. The staggering profits of crude oil have launched select governments and corporations into a new stratosphere of influence as the vested interests of oil charter a shadow empire defining exactly where the prerogatives of democracy, sovereignty, and even war must cease and desist. For fans and critics alike, the attributes of the commodity must ground any serious understanding of crude oil. Only occasionally is that ground sullied by events that disrupt the momentum of gain, but these disasters are always exceptions to the rule. The real story of crude oil begins and ends in the commodity form.

This book advances an inquiry in the opposite direction. Crude oil is the world's greatest disaster that only for the briefest of moments coheres in the commodity form. Destruction is the norm; the commodity is the event. Long apparent to residents of extractive frontiers and frontline communities, the ascent of the oil industry is a story of destruction rippling ever outward. This outlook, against a widening backdrop of damaged life and planetary systems tipping into disarray, is becoming both more scientific and more commonplace. In rising seas and dead zones, in cancer clusters and superstorms, in rapacious histories and foreclosed futures, the disasters of crude oil routinely exceed any register of reasonable gain. While logics of accumulation buttress the business of fossil fuels and remain integral to their history, the cellular, social, and earthly disruptions unleashed by such a business grossly exceed the analytics of capital. In stunted forests, obstructed migratory routes, asphyxiated ocean layers, deformed animals, poisonous groundwater, and contorted elemental cycles, the negative ecologies of crude oil outweigh any accrual of profit or power.

Fossil fuels are destroying the world. Scientific estimations of the impact our current consumption of crude oil is having outpace the most imaginative savants of apocalypse: entire landscapes rendered inhospitable, superstorms beyond our ability to withstand, wild infernos beyond our ability to extinguish, roving droughts in prime agricultural land, glaciers and permafrost awash in torrents of melt, flooding of the densely populated shoreline, entire oceans becoming acidic beyond the window of most marine life, and extinctions on a par with a meteor strike. As they have for generations, the world's poor will bear the brunt of these impacts. Yet the impact cannot be contained within the given structures of inequality. What happens after the commodity now threatens the biochemical and meteorological conditions of life itself.

Such destruction sets the stage upon which this book proceeds. From leaky refineries to extractive frontiers to contaminated landscapes, this book describes the manifold disasters of crude oil. Yet the central aim of this book is less documenting such destruction than examining the ways we've grown accustomed to cordoning off these casualties as a secondary matter of concern, the ways our apprehension of such destruction presumes something already within our ability to remedy. Whether in the environmental monitoring networks that encircle drilling pads or petrochemical plants or in the environmental planning documents that encase pipeline projects and drilling leases, the disastrous properties of the oil industry are rarely apprehended directly. Rather, they are first distilled into select representative problems, each atomized threat thoroughly fenced off in administrated safeguards, engineered moderation, and the feeling of well-managed risk. Together, these protections pull select harms of the oil industry into a separate ledger while turning a blind eye to the wider fields of devastation underway. These protections can be quite effective at tempering destruction, and the deep investments in their format often lead many advocacy campaigns and protests to ally their complaints with the official accounting of harm. Such protections also allow the fiscal properties of the oil to ascend into the commodity, as if weightless.

The sundering of material harm from material gain is a work ongoing, a conceptual dissolution that is fully integrated into the design, function, and regulation of nearly every hydrocarbon installation worldwide. It is also fundamental to response efforts when something goes wrong. This work ongoing forms the central field site of this book, both in the "environmental crisis" of the 1960s that first catapulted the catastrophic impacts of fossil fuels into public prominence in the United States and the legislative struggles that then ranked those injuries as secondary to the energy economy, and in the contemporary institutional fields, scientific norms, and ethical tactics that work to reinstate this pernicious premise over and again: the reason of the commodity exceeds the reach of damaged life. Such an outlook saturates the infrastructure of oil, where loss is always a line item within the ledger of gain. Within extractive operations, regulatory actions, activist demands, and courtroom settlements, the damages of oil are often objectified by first placing them in subservient relation to towering profits. It has become common sense. No injury is possible beyond the market valuation of crude oil, no damage is credible beyond the ability of profit to compensate, no disaster is plausible beyond the technical capacity of the industry to contain its spread. In the bustling business of oil, additions are always greater than subtractions.

The opposite equation haunts the landscapes and livelihoods injured by oil. The laments of those living near drilling sites or refineries or pipeline projects trip up the professional taming of risk and pedigreed hierarchies of profit and loss. Privileging such dissent enables ethnography to explore the tremendous categorical labor and scientific infrastructure required to render the open-ended assault of the oil industry as a set of discrete and ordinary technical challenges, as something perpetually provincial to the tremendous wealth pouring forth. Jagged and unvetted as they are, these jeremiads form the methodological and theoretical opening of this book. Such an orientation finds solid anthropological anchor in the righteous outrage of frontline communities beset by asphyxiating refinery emissions, of farmers realizing a nearby petrochemical plant has contaminated soil and groundwater tended for generations, of vibrant hunting and fishing communities suddenly severed from natural abundance by the poisonous runoff from oil drilling sites, of factory towns learning the carcinogenic residue of plastics

FIGURE 1. Haul road along the Trans-Alaska Pipeline. Photo by author.

manufacturing has been in the public drinking water for decades, of coastal communities devastated by oil spills and encroaching seas, and of residents reckoning with how the toxic fallout of fossil fuels is salting the corner of this earth they know best. Centering such outrage, this book refuses the prevailing premise that the ecological fallout of the oil industry is sensibly contained within the imperial logic of profits, that the destruction unleashed by oil extraction, refining, and combustion is always already dislodged from and deferential to its fiscal promise. Against a pervading logic of positive surplus, this book revolves around the negative excess of the oil industry. With a commitment to practical justice throughout, this book asks what comes into view when you allow for injury beyond immediate remedy? What revolutionary alliances of science and protest stir in the centering of material fields of devastation that exceed any official accounting?

Karl Marx once called the commodity the first "citizen of the world," a pioneer in its ability to chart a world without borders. The rise of the commodity disciplined the world into a shared rationality of exchange, offered metaphysical endorsement of the upsurge in inequality, and introduced exploited labor as a more potent empirical basis for universal dissent. Today, the widespread resonance of ecological collapse charts out another landscape of common cause in the evening shadow of our most exemplary commodity. While those profiting from the oil industry grow ever more rarified and removed, the toxic disruptions and climactic fallout of fossil fuels seep into every corner of the planet. We inhabit a contemporary moment of parched cities and scorched forests, desolate reefs and inundated coasts, contaminated neighborhoods and poisoned warfare, withered farms and ransacked landscapes, of the conditions of life coming undone as turbo-fueled accumulation resurrects ancient questions of basic survival. In the disastrous wake of the oil industry, the commodity may no longer be our elementary form so much as our terminal diagnosis. While anthropology has much to say about how our synthetic ecological collapse divides us even more forcefully into the hierarchies that first made the commodity possible, it has much less to say about how the ends of the commodity are already drawing us together in new ways.

THE DISASTROUS HISTORY OF THE ENVIRONMENT

If untamed destruction sets the theoretical stage of this book, the environment takes the leading ethnographic role. The ordinary work of the environment is at the core of this inquiry, both in the disruptions it brings into technical coherence and the experience of disruption it evacuates in so doing. In design and in practice, the environment has an almost magical ability to tame the negative ecologies of crude oil and flatten earthly laments into mere superstition. Yet not only does the environment discipline destruction, such destruction may very well be the perennial wellspring of the environment itself.

This book describes the outsized role fossil fuels have played in making the environment visible, factual, and politically operable. This book draws historical and ethnographic attention not only to *what* we know of the environment but also to *how* we have come to know the environment. To a striking degree, the specific crisis the environment realizes, the forms of understanding and responsibility it authorizes, and the horizons of action and anticipation it routinizes all bear the imprint of destructive hydrocarbon afterlives. Whether by way of urban smog or petrochemical runoff or drilling frontiers or even oil spills, as fossil fuels unravel the conditions of life, they also instigate new authorities to monitor and police those conditions. Yet the resulting definition of the defendable environment, wedged in between hydrocarbon pollution and public outrage, has often been effective to the extent that it sidesteps the underlying petro-problems and focuses attention instead on stabilizing the mediums of exposure, like clean air and clean water, and perhaps now a stable climate. Not only does the environment divorce measures of harm from measures of gain, but the category itself comes to configure the moderate contamination of life as completely natural and the incidents of verified harm as secondary to the fueled progress underway. Today, as the disruptions of fossil fuels snap back into focus around rising planetary concerns like global warming, ocean acidification, and the Anthropocene, hydrocarbons can appear as an unprecedented crisis bearing down on the present. This book documents the impacts, injuries, and disasters that have long accompanied fossil fuels and the manner in which our solutions have often been less about confronting the cause than managing the effects. This history of our present promises to resituate scholarly understandings of fossil fuels and renovate environmental critique today.

This book builds on a decade of anthropological research on hydrocarbon projects and problems in North America. For most of the past century, the United States has been the world's largest producer, refiner, and consumer of fossil fuels and petrochemicals. As such, the imperial energy networks and scarred landscapes of the United States are central to this question of the environment. The book opens with a brief history of the category of the environment, showing how rising hydrocarbon pollution and petrochemical fallout convinced scientists, citizens, and policymakers that a new concept was needed to face up to the disconcerting breadth and intimacy of this new problem. As two leading public health officials wrote in 1955, the advent of the "the era of synthetics and the petroleum economy, when combined with epidemiological observations, indicate that a general population hazard is of more than theoretical significance" (Kotin and Hueper 1955: 331). It was around these concerns that the environment first found effective definition in the United States and soon came to monopolize popular and scientific understandings of damaged life and the state's obligation to it.

In the late 1960s and early 1970s, the environment shifted from an erudite emphasis on the influence of context to the premier diagnostic of a new world of manufactured precarity. As shorthand for the resulting crisis of life, the environment became an insurgent field devoted to understanding disrupted life and taking responsibility for it. The environment, a term "once so infrequent and now becoming so universal," as the director of the Nature Conservatory commented in 1973 (Nicholson 1970: 5), soon found official recognition in new agencies and ministries in governments across the globe. Pointing out the shortcomings of the nature/culture dualism long before such a thing was fashionable, the resulting constitution of the environment pulled earthly mediums into national governance, foregrounded survival over nostalgia, moved past a politics of purity, and acknowledged a world alive beyond our conception of it. If the rise of the environment previews these contemporary themes, its history also carries a warning: the growing recognition of the crisis of life paradoxically narrowed the grounds of effective critique within it.

Reviewing the history of the category, this book also explores two possibilities of the environment that almost came to be: a more serious grappling with negative ecologies in the work of Rachel Carson, Barry Commoner, and others, and a brief moment at the UN Conference on the Human Environment in 1972 that insisted the ecological debts of colonialism be weighed alongside a "world-wide harmonization of standards" in establishing planetary environmental protections (UN 1973: 26). Each of these possibilities was blunted by the rise of two techniques that now instantiate the environment: toxic thresholds and impact assessments. In different ways, these techniques respond to the disastrous materiality of fossil fuels, and each functions by turning the injurious reach of hydrocarbons into a kind of field laboratory for the measurement and management of endangered life. Toxic thresholds and impact assessments both work to locate injuries within the register of corporate feasibility.

Toxic thresholds have been extraordinarily effective at reining in air and water pollution within their jurisdiction (there is some evidence that they made things worse for those just outside such jurisdictions). Thresholds work. But what work do they do? First of all, thresholds authorize pollution, to a point. In 1958, ecologist Paul Shepard complained that thresholds "idealize life with only its head out of water, inches above the limits of toleration. [. . .] Who would want to live in a world which is just not quite fatal?" (395). Noting the "concessional character" of thresholds, Ulrich Beck (1993: 64–65) more recently described them as tools that, while they "may prevent the worst," nonetheless should be seen as authorizing "the permissible extent of poisoning." Second, thresholds turn toxicity into an event. Thresholds are the condition of possibility for toxic events. Harm is no longer a fundamental property of certain processes or products, like extractive operations or petrochemicals, but an exceptional event, a fleeting density in time and space. Thresholds turn attention away from material structures of harm and toward momentary ruptures in the official definition of harm. Thresholds transform the extremities of harm into the only thing that matters. Finally, thresholds erase the embedded and embodied experience of toxicity. They carry a "body-blindness," as Christopher Sellers (1999: 58) has put it. Thresholds build up an infrastructure of concern that displaces the "bodily archive" of lived toxic exposures in favor of abstract and discrete deviations from implemented norms (Brown 2016: 46). This not only sidesteps the colluding ecologies of toxicity that assail certain neighborhoods and certain landscapes, it also means the environment, by design, is unable to register the historical inflections of class, race, and gender so often wrapped up in the toxic problems it purports to address.

Impact assessments compel the fragility of life into decision-making. Introducing the conditions of life to decision-making has been hugely influential, yet impact assessments often work in unexpected ways. To the frustration of many citizens who participate in environmental impact assessments, voiced concerns are not akin to voting on a potential project. While environmental impact assessment meetings can provide a microphone for lived and livid concerns, all too frequently they do so only to deny those voices any means of amplifying themselves into a more transformative politics. They don't so much refute critique as exhaust it. How? First off, impact assessments acknowledge detrimental impacts, only to co-op them. By claiming the perspective of potential harm, environmental impact assessments internalize what had previously been an external position of critique. Critique is drained of the capacity to confront extractive projects; it is instead drafted into an unpaid position in the very design and operation of those extractions. This process is often marked by an engineering hubris that believes every potential impact can be mitigated and managed with the right combination of planning and technology. Environmental impact assessment, then, may be one of the ethical stances that enliven contemporary capitalism, as suggested by Boltanski and Chiapello (2007). Second, impact assessments map the limits of legibility (Checker 2007). As Andrew Barry (2013) has shown, by making the potential impact of a project visible, environmental impact assessments "mark out—however provisionally—the limits of [of a company's] social and environmental responsibility" (19). Finally, impact assessments reify the particularity of a place, not through history or ethnography but in abstract relation to the footprint of a project. Environmental impact assessments extract the moment just before disruption and project it as an authoritative definition of normal life. Such definitions erase chronologies of change, becoming inflexible measures that dictate the legitimacy of subsequent discontent and suffering. As

I have written elsewhere with Lucas Bessire, such work "narrows the areas of legitimate concern and widens the scope of acceptable disregard" (Bessire & Bond 2014: 441).

Toxic thresholds and impact assessments are not neutral scientific innovations. The historical development of both toxic thresholds and impact assessments is deeply tied up with the oil industry. In various ways, each was developed in technocratic efforts to rein in the destructive reach of fossil fuels without disrupting their profitability. Toxic thresholds and impact assessments can be ruthlessly proficient, and instantiations of the environment along these lines have been instrumental in not only authorizing entirely new fields of science and law but also saving lives and reducing pollution worldwide. Yet the resulting definition of the environment has often been effective to the extent that it sidesteps the underlying cause of the problem—the oil industry—and focuses attention instead on stabilizing the mediums of exposure and reifying the moral boundaries of their operations. In so doing, thresholds and impact assessments cleave matters of harm from matters of gain. Managing the degenerative effects of fossil fuels becomes an autonomous field of research and regulation, a separate and secondary matter of concern.

Toxic thresholds and impact assessments also do crucial normative work. Thresholds and assessments establish the normative criteria for environmental critique in science, law, and advocacy. But here widespread agreement on the normative basis of critique does not open up the possibility of a more transformative politics, it forecloses it (contra Habermas). Displacing a politics of confrontation, toxic thresholds and impact assessments push effective action into the realm of standardized methods, certified results, acceptable levels, and codified assessment models. Quietly orienting the state's forceful considerations as well as its averted gazes, thresholds and impact assessments have become vital normative technologies within contemporary politics. Instantiating the official definition of defendable life, such scientific and legal norms also introduce a technical limit to democratic practice around fossil fuels. Overriding any popular consensus about the source of harm and what might be done, the primary lever of state attention instead shifts to certified deviations from the norm. The environment, here, comes into historical and ethnographic focus not as the answer to the crisis of life engendered by fossil fuels but as a way of governing the resulting contradictions. It is no coincidence that the rise of the environment mirrors the unbound consumption of fossil fuels in the United States.

And it is no coincidence that the environment has not so much checked our addiction to fossil fuels as provided acceptable parameters for that addiction to deepen and expand.

FIELDWORK IN DISSONANCE

It was somewhere along the Gulf Coast at the height of the BP oil spill in 2010 that it first struck me, the tremendous dissonance between how the state spoke about the disaster and how the residents experienced it. I was following a caravan of federal officials as they drove from city to city. Each evening the same information booths would be set up in different high school gymnasiums and the same PowerPoint presentation would explain the deepwater blowout to a new coastal community and how its impact was already being resolved. Afterward, when asked if they had any questions, residents would find their way to the microphone and talk about how the oil spill was reaching into their bodies, stealing what little stability they had built up, and upending their lives. State officials never quite knew how to respond. They would thank residents for sharing before reiterating plans for various studies of marine life, millions of dollars in wetland restoration, and new public access points to the shoreline. Almost uniformly, these plans had nothing to do with how residents experienced the disaster. For state officials, the oil spill was a reasonable problem suited to environmental governance, a momentary rupture easily amended by dipping into the perennial fortunes of the industry. And for those who knew the script and how to place their own agendas within it, the oil spill proved quite the boon. For many residents, however, the oil spill veered past the edges of technical reason. Their ragged experiences refused the instruments of feasibility, opening a wound that reached past available measures of injury and recompense. In these encounters, the promise of fueled progress that underlies so much of the contemporary world fell to the wayside as the balanced architecture of profit and loss came undone.

I've witnessed variations of this scene all across North America, in polished industry campaigns for environmental stewardship in offshore developments along the Gulf Coast and in the museum of tar sands of Alberta, in routine impact assessment meetings in midwestern towns along the Keystone XL route and in Alaskan villages adjacent to wildlife refuges opened for drilling, and in angry information sessions about the risks of living near refineries in the Caribbean or plastics factories in New England now saturated with petrochemicals. Again and again the encounter is repeated: company representatives and state officials describe the problems of oil as firmly within the enhanced environmental capacities of the industry and the state, while nearby residents voice a destruction pulling life away from any practical criterion of control. This ethnographic dissonance forms the theoretical stance of this book.

Such fraught scenes are far from novel, and highlighting their longer history in policy and scholarship forms a key part of this inquiry. Yet in contemporary scholarship, the disastrous reach of the oil industry is often understood in one of two ways: by either taking up the outlook of the state as the normative basis of justice or anchoring critique to a conceptual horizon beyond present destruction. Environmental justice scholarship often works to bend the jarring scenes described here into official legibility, while ontological scholarship often strives to more fully inhabit the theoretical redemption of such inharmonious scenes. Both aim to resolve the dissonance of petro-destruction, one through juridical means and the other through conceptual means. There are many merits to both approaches: the former builds an effective moral and legal case against the oil industry, while the latter crafts new tools of critical renewal not already complicit in the ransacking of the planet. Environmental justice scholarship advances real change within the system, helping pull long-standing injuries into irrefutable claims for stately recognition and desperately needed compensation. Ontological approaches refuse the system entirely, and by locating alternative grounds for theory such approaches can help realize the possibility of a world beyond profit and power into being. And while each has added crucial insight to the scholarly critique of the oil industry, neither seems sufficient to the crucible of now. Environmental justice scholarship struggles to contest the conceptual architecture of profit and loss that underlies the oil economy, while ontological scholarship struggles to advance practical justice in the present-tense.

Rather than immediately trying to resolve the material incongruity of crude oil, this book stays close to it. The dissonance between the promise of oil and its ecological unrest is in itself immensely generative, for environmental science and policy no less than for ethnography. Scholarship that too quickly moves to resolve such dissonance can miss the governing institutions, analytical technologies, and corporate investments working in the same direction. Theoretically pausing in scenes of dissonance brings this ordering work of the environment into stark ethnographic focus.

Centering this book on the disastrous excesses of the oil industry displaces the commodity as its definitive form and opens up new ground for ethnographic critique. Foregrounding the negative ecologies of crude oil and the social dissonance around them provincializes the overbearing logic of gain without ever leaving the scene of its crime. It provides a way to neither innocently inhabit nor wholly refuse the official accounting of harm but instead attend to the tremendous epistemic and infrastructural labor involved in disciplining broken worlds into such accounting, as well as seeing what does not add up. Ethnographic attention to negative ecologies is not a form of theoretical despair and even less a call for political resignation in the face of overwhelming destruction. Negative ecologies brought me closer to the battered world at hand with an aim to do something about it. Ethnographic attention to negative ecologies allowed me to see close up all of the scientific work being done to reorder the world without either taking up the teleology of that work or refusing its significance entirely. The borderlands between mastery and destruction are prolific, for regulatory science no less than for political refusal. Not only are the negative ecologies of crude oil at the forefront of innovations in environmental science and policy, but their growing recognition leads many frontline communities to refuse offers of managed risk and instead stand more forcefully against the oil industry itself. Allowing destruction to reach beyond reason of the commodity, this book joins with these protests to advance a critique aiming to dismantle the oil economy from within the conceptual grounds of its operations while also advancing the urgent claims of those injured by it.

From extractive frontiers in Canada to entrepôt refineries in the Caribbean, from oil spills to the toxic fallout of plastics manufacturing, these chapters describe the scientific and ethical work that is disciplining these worlds into the legibility of the environment as well as how nearby communities come to live within and against toxic thresholds and impact assessments. Each of these communities is more than an intellectual curiosity. At each, I was drawn into the struggles of nearby residents against fossil fuels, sometimes volunteering my time with existing organizations and sometimes playing a more active role in tactical pursuits of justice. Some of these sites were visited for short periods; others involved extended fieldwork. One of the sites is my current home, where petrochemical carcinogens were discovered to have extensively contaminated the region's soil and groundwater in 2016, including my own backyard.

The theoretical arc of this book first took shape in my growing ethnographic sensitivity to the operational gravity of the environment. Talking with those living on the destructive edges of the oil industry and with those scientists and agencies tasked with managing such zones, I became attentive to the ways in which the terms and technologies of the environment pulled shocking harm into the form of a reasonable problem, smoothed out the jarring edges of angry residents and wounded landscapes, and measured what seemed to defy measurement (while discarding what didn't fit). In some of these engagements I found myself aligned with the environment, demanding lower thresholds or better impact assessments, arguing with disembodied facts, or even inhabiting the bureaucratic procedures of the environment with the hopes of slightly enlarging their reach. In other engagements, I found myself refusing the logics of thresholds and impact assessments, demanding recognition of what Kate Brown (2016: 46) calls the "bodily archive" of contamination and loudly criticizing the institutional limits of the environment. This book, in many ways, is my attempt to gather together these experiences to reflect more clearly on what work the environment does, how it came to exert such influence over us, and how we might reclaim its founding prompt while shedding its more complicit forms. The chapters move progressively from cases in which toxic thresholds and impact assessments effectively cordoned off the disastrous properties of fossil fuels to cases in which the negative ecologies of crude oil overwhelm thresholds and assessments, suggesting alternative scientific and political arrangements.

ORGANIZATION OF THE BOOK

Environment

The environment often seems far too easy, far too obligatory, and far too footloose a concept to warrant serious attention. In the rush to move past the environment, few have attended to the history of the concept. This chapter brings new attention to this neglected history. In the late 1960s and early 1970s, the environment shifted from an erudite shorthand for the influence of context to the premier diagnostic of a troubling new world of induced precarity (whether called *Umwelt*, *l'environnement*, *medio ambiente*, *huanjing*, *mazingira*, or *lingkungan*). While the resulting recognition of the environment largely consisted of bringing existing problems together under one umbrella—factory pollution, urban sewage, radioactive fallout, automobile emissions, garbage disposal, and even global warming—the resulting synthesis was powerful. As shorthand for the resulting crisis of life, the environment became an insurgent field devoted to understanding contaminated life and taking responsibility for it. This introductory chapter traces two techniques that work to instantiate the environment: toxic thresholds and impact assessments. In different ways, each responds to the disastrous materiality of fossil fuels, and each functions by turning the injurious reach of hydrocarbons into a kind of field laboratory for the standardized measurement and management of endangered life. Toxic thresholds were nationalized in the United States around the emissions of hydrocarbon combustion and perversely provide a novel means of authorizing emissions up to the exact point of harm. Displacing efforts to expand environmental rights, impact assessments instead brought scientific attention to stabilizing the conditions of life. By many accounts, the fossil fuel industry is the most heavily invested industry in impact assessment. Quietly orienting the state's forceful considerations as well as its averted gazes, thresholds and impact assessments became both a vital object of contemporary politics and a technical limit to democratic practice. My sketch of this history is less a finished project than a preliminary effort to brush away the accredited nonsense clamoring to contain the frontline laments of contamination. Here, I aim to provide a much deeper historical and theoretical credence to their complaints. Such work is not aimed at getting away from the present but at providing new coordinates for ethnography to come closer still to the ecological crisis of now.

Governing Disaster

One of the hardest things about studying the largest oil spill in US history was finding it. On the ground, the BP oil spill was not always obvious. A television crew I met had been on the Gulf Coast for days looking for the disaster. All they found were tar balls and anecdotes, nothing spectacular. "Where is the spill?" they asked me. This chapter describes how scientists and federal officials scrambled to bring the unprecedented properties of this sprawling deepwater blowout into analytical and regulatory focus. Embedded in university laboratories and emergency response teams, in chapter 2 I describe how the BP oil spill went from a sprawling mess into a manageable problem by first transforming the ocean into a sort of scientific laboratory. For marine scientists who responded to this unprecedented deepwater blowout, making sense of the disaster rested on first stabilizing an unwieldy field of inquiry, codifying novel analytical techniques, and then cultivating a new ethos of scientific objectivity under duress. The disaster did not begin as a clearly defined event; it became one through this tremendous labor of science and technology. For one, techniques of measuring the oil spill instigated a new understanding of the baseline conditions of life in the Gulf of Mexico and posed new questions about our political, scientific, and ethical relationship to that baseline. That is to say, the oil spill materialized a new version of the environment itself. Here, the newly coordinated environment not only objectified the oil spill, it also quietly defined what knowledge of the disaster and what relations to it could have credibility. The environment fully contained the disaster, insulating the biological reach of this oil spill from human consideration and rendering personal accounts of sickness implausible and illegible. Techniques of sequestering and inspecting the oil spill came to underwrite a new regime of disconnection between the disaster and the public.

Ethical Oil

One of the more remarkable things about the tar sands in Alberta is how upfront oil companies are about their impact on the landscape. On local billboards and in interviews, tar sands operators regularly acknowledge that they are going to destroy the place. After all, they say, this is "the real cost of energy today." Such acknowledgments, however, quickly pivot toward the huge investments the industry is making in its ability to put it all back together again. With restoration projects that strive to join the best of environmental science and the most traditional Indigenous ways of life, oil companies proudly tout their unique ability to engineer a more culturally informed boreal forest, to build a more cosmologically attuned ecosystem. Drawing on ethnographic research on corporate social responsibility in the tar sands of Alberta, chapter 3 describes how energy companies objectify the Indigenous environment to selectively manage the impacts of tar sand extraction and to redeem their ongoing petro-destruction from the perspective of a premodern future. Here, the Indigenous environment comes into focus neither as the dated epistemology of anthropology nor as the ontological footing of radical alterity but rather as a novel moral technology for redistributing hydrocarbon risks and responsibilities in space and time.

Occupying the Implication

Chapter 4 presents a pivot of the book. Before this chapter, the negative excess of fossil fuels is described as a frontier of state power and corporate responsibility. Tracking pipeline protests, this chapter shifts the direction of inquiry to how the negative excess is opening new fields of contestation and demand. Today, pipeline protests have become a common occurrence. Whether in national flashpoints like Keystone XL and Dakota Access Pipeline or in more regional protests around the Addison Natural Gas Pipeline in Vermont and Kinder Morgan Pipeline in Massachusetts, the rather bureaucratic process of permitting a new pipeline has opened up a formal venue to air grievances over histories of extraction and the foreclosed futures they herald. These protests not only puncture scholarly arguments about how petroleum infrastructure evades democratic accountability (Mitchell 2011), but they also have become a front line in a conflict so immense that many have struggled to sort out where to begin: the fight to hold back the worst of climate change. This chapter focuses attention on the formulaic process that sets the stage for so many contemporary pipeline protests: environmental impact assessments. Planetary crisis, in these sites, comes to matter as both the inevitable endpoint of destructive logics and the spark that might ignite a more radical accounting. This reckoning is philosophically attuned to the impact assessment process, even as the demands being made wildly exceed its legal and institutional capacity. If such protests rarely seize control of the pipelines, they nonetheless seize the implications.

Petrochemical Fallout

In 2014, the chemical perfluorooctanoic acid (C8 or PFOA) was discovered in the public drinking water in the Village of Hoosick Falls, New York, and soon after in residential wells around Hoosick Falls, New York, and Bennington, Vermont. Once a key ingredient in the manufacture of high-performance plastics like Gore-Tex and nonstick kitchenware, PFOA is a synthetic petrochemical that is persistent, mobile, and toxic. Synthesized by American chemical manufacturers in the 1940s before being banned in the United States in 2015, PFOA has spread worldwide, is durable on the order of centuries, and is now found in the bodies of most living creatures on Earth. Yet the near-universal reach of the problem is at odds with the lived dimensions of it: the experience of PFOA toxicity remains confined to drinking water disasters near plastic manufacturing hubs in the United States. Chapter 5, drawing on three years of involvement with the discovery of PFOA in my adopted hometown, asks how anthropology might recognize the haunted landscapes of petrotoxicity in conversation with the embodied experience of them. In doing so, this chapter reflects on how ethnography can take stock of contamination without getting swept up in the evocative purchase of the term today. Against recent theorizations of contamination as a kind of emancipatory release from modern categories, this chapter follows three communities as they struggle to secure clean drinking water and grapple with a legacy of contamination that routinely exceeds the available levers of environmental justice.

The Ecological Mangrove

Chapter 6 links up the colonial history of fossil fuels with the celebrated ecology of mangroves in the Caribbean. Building on ethnographic and historical research in Puerto Rico and St. Croix, it outlines the often neglected but quite consequential place of crude oil in the Caribbean. After discussing the construction of what became the second largest refinery in the world, I describe how the imperial energy networks of the United States first came to the Caribbean. Troubling a popular origin story of the Caribbean, colonial and industry leaders voiced a robust critique of the plantation in order to justify the introduction of these enclave refineries. Imperial energy networks welcomed an unprecedented problem to the region: coastal oil spills. The scientific and legal response to these spills first worked to provisionally extend the logic and operations of the environment to this colonial setting. Yet what they discovered soon exceeded those forms: namely, the vital relationality of mangroves. Rather perversely, the destruction of coastal ecosystems in the Caribbean—in which crude oil played the leading role—opened mangroves up to new forms of knowledge and care. While many claim that fossil fuels helped cultivate a modern disregard for the natural world, I show how the negative ecologies of fossil fuels also instigated new scientific and political appreciations for the liveliness of the natural world. And soon the figure of the mangrove was used to ground political possibilities that not only advanced new critiques of the empire of oil in the region but also envisioned new ways of living beyond it.

. . .

These chapters stay close to the slow violence that coheres around hydrocarbon infrastructure in the United States and the fields of calculation and care that both recognize such violence only to abstract it from social history. While the United States figures prominently in this account, my book notes the circulations and conjectures that layered this concept of the environment onto other nation-states, whether by exporting the US model as a stipulation of developmental funds, as the dawning of commensurable problems like radioactivity that borrowed from existing expertise, or as the preferred method for extractive industries to delineate narrow responsibility for harm while authorizing intensified production. Moreover, my research and writing presume that many of the social movements that have cohered around the environment in a variety of local, national, and international contexts (e.g., Guha 2000; Martinez-Alier 2002; Radkau 2014) have long tripped up and exceeded the technical constitution of the category, even as the state-backed objectification of the environment has animated the analytical and ethical justification of those movements.

Attentive to the material dissonance of the oil industry, this book is less an ethnography of a single place than of a problem shared across hydrocarbon frontiers in North America. As these chapters demonstrate, negative ecologies offer a different way to ethnographically apprehend and theoretically assail the oil industry today.

THE NEGATIVE ECOLOGIES OF CRUDE OIL

Today, many critical scholars focus new attention to how the physical properties of fossil fuels helped consolidate some of the most consequential social forms of our present: from the state (Coronil 1997) to transnational corporations (Coll 2012), from neoliberalism (Harvey 2007) to the economy itself (Mitchell 2011). While this scholarship is instructive, the main thrust of its critique rests on linking the material force of hydrocarbons to positive iterations of capital and power. This book turns the question of the materiality of fossil fuels in a different direction, reflecting on how the negative ecologies of fossil fuels draw injured webs of life into new empirical and ethical focus. Here, the ongoing substitution of hydrocarbon efficacy for coerced labor offers a tragic twist to the analytics of historical materialism. The telling tensions of our contemporary era are not only ordered along the contradictions of accumulation or power but also in accordance with the contractions of life, whether in rising rates of cancer or rising levels of seawater. Every human on Earth and most of the animals now host petrochemicals and other molecular traces of the fossil economy in their bodies. This "wonder world" of hydrocarbon innovation, as Rachel Carson (1962: 11) once called it, now heralds uneven worlds of broken landscapes, asthmatic populations, dead zones, and distorted atmospheric systems.

From the afflictions of frontline communities to the collapsing thermal boundaries of the planet itself, the destructive reach of the oil industry haunts the limits of mastery. When the Caribbean was swept up in the promise of becoming the world's premier oil refining hub in the 1970s, the destruction of mangroves in the construction of hydrocarbon ports was widely celebrated as evidence of modern progress. While oil refining has largely come and gone in the Caribbean, the decimation of coastal mangroves continues to ripple outward in shoreline erosion, storm surges, and still collapsing fisheries. In St. Croix, which was home to the largest refinery and petrochemical plant in the world for a number of years, the refinery laid waste to the island's freshwater aquifer and rendered large parts of the island unsuitable for farming or housing before filing for bankruptcy at the first sign of a real investigation into these damages. And then another aftershock of oil arrived: superstorms. In 2017, two back-to-back category 5 hurricanes slammed into St. Croix, battering an already broken island.

Or consider the toxic fallout of plastics. Engineered to be impervious to every natural degradation process, the petrochemical PFOA was used extensively in plastics manufacturing and emitted in shockingly large amounts without regard to what happened next. Even though these chemicals never break down, their geological stability is far from benign. Despite immense corporate efforts to bury the evidence, trace exposure to these chemicals has been sickening workers in plastics plants for decades and now assails the health of communities around plastics factories. First synthesized by American petrochemical companies in the 1940s, these "forever chemicals" now blanket entire landscapes and are found in most cellular forms of life on Earth, introducing a scale of petrochemical contamination far beyond our capacity to remediate even as we learn more about the harms already underway.

Fossil fuels, in this reckoning, appear as a haunting gift, and perhaps it is Marcel Mauss more than Marx who offers the most exacting conceptualization of the materiality of fossil fuels. Fossil fuels are a negative gift. When consumed, hydrocarbons do not disappear but can come to structure relations of obligation that may exceed the capacities of life itself. Yet this destruction carries its own rippling creativity. In sharp and subtle ways, as fossil fuels assail the underlying relationality of life, they open those vital ecologies to new forms of understanding and responsibility. While the rising prominence of the Anthropocene solicits rapt attention on the impending foreclosures heralded by hydrocarbon emissions, such a project frequently sidesteps the longer history of acknowledging and managing the disastrous qualities of fossil fuels (Bonneuil and Fressoz 2016). The disasters of oil are more than a looming catastrophe; they are also a fractured history of our present, the "shadow kingdom" that haunts the modernist order of contemporary, as Ulrich Beck (1993: 72) described it. These concerns not only offer a crucial correction to our understanding of fossil fuels, they also offer a generative site to engage new insights on vital materiality and to reflect on the often-obscured relation of nonhuman agency to questions of social justice and critique.

Ethnographic attention to the negative ecologies of the oil industry provides a novel entry point into the changing status of materiality in anthropology today. So many of the dizzying reformations of materiality unfolding in anthropology emerge from incisive ethnographic encounters: laboratory science, built infrastructure, Indigenous cosmologies, multispecies collaborations, and feral ecologies, to name a few. All of these crafted new sensitivity to the capacities of other species, landscapes, and technologies to shape the world at hand (even as such profound influence was rendered illegible by enlightened frameworks of the real). Analytically attuning to these more-than-human capacities provided scholars a fertile place to begin anew, an insight that almost seemed capable of giving birth to a new world already latent within our own.

Much of this revival of materiality proceeds by allying ethnographic curiosity, political practice, and critical theory with the underappreciated agencies of the natural world. Collaboration with the effervescent physicality of plants and animals, rivers and mountains becomes a more potent form of critique, at once derailing the tyranny of humanist understandings and lifting alternative arrangements into new significance as theory and template. Whether by cyborgs, domesticated microbes, companion species, channeled rivers, or cacophonous rain forests, it is striking how many of the field sites that are revitalizing materiality in scholarship stay firmly within the positive attributes of the material world. With strikingly few exceptions (Farmer 2001; Masco 2006; Stoler 2013; Fortun 2014; Gordillo 2014; Bessire 2021; Khayyat 2022), ethnographic encounters with the negative ecologies of the contemporary have been held at arm's length in the more pronounced theoretical reformation of materiality in anthropology. Part of this, I think, has to do with how this renewed optimism of the physical sits uneasily with a growing pessimism of the political. Against the colonial complicity of liberal democracy and the complete failure of modernist epistemologies to face up to the worsening condition of most, autochthonous ecologies promise an empirically rich ground to fundamentally break with the irredeemable history of the present in order to chart out more equitable ways of living together. As methodological tactic, concrete heuristic, and incipient revolution, ecology is drawn into ethnographic significance by the positive theoretical vision of the world to come that it advances. Wedding anthological curiosity and critique to these heterodox ecologies becomes both prophecy and proof that a better world is possible. What about negative ecologies?

Whether in the destructive afterlives of nuclear weapons, land mines, toxic dumps, fossil fuels, or petrochemicals, some feral material is not so easily channeled or co-opted. Nor does the oil industry—and its ransacking of the planet—seem particularly troubled by theories of an ecological otherwise. Indeed, as I show about the tar sands, such heterodox ecologies are now at the forefront of corporate remediation plans in a manner that authorizes intensified extraction under the promise finally allowing landscapes to align with more radical ecologies once the taint of oil is removed. Negative ecologies open the door for ethnography to grapple with the destructive legacy of modern progress without first domesticating it into complicit institutions or theoretically departing from its contingent history. And it is from such a contingent history that justice in the present tense remains possible without fully endorsing the colluding techniques that have sprouted up to manage the problems without ever solving them. Such a stance—at once analytical and political—insists on practical assistance for those whose lives have been upturned by fossil fuels without ever losing sight of the revolutionary break with fossil fuels that alone will make a better world possible. Negative ecologies turn ethnographic attention back to materialism, but less as a utopian departure from history than as an effective way to trip up the conceit of that history from within the inhabited ground of its operations. In so doing, this book hews close to what so many living on the front lines of the fossil fuel industry have long known: real justice necessarily involves toppling the system that renders their lives secondary to the fiscal fantasies of fossil fuels.

In conversation with the anthropology of science, questions of vital materiality, and new social research on fossil fuels, this book describes how the deleterious afterlives of fossil fuels gain scientific definition and to what political effect. This book is particularly attentive to the ways in which (1) the material force of fossil fuels is not fully expended in the moment of combustion but often comes to haunt life with socio-chemical traces and attritional violence; (2) environmental protections do not precede the disasters of fossil fuels but often emerge from them; (3) the objectification of the environment can gloss over embedded and embodied histories of harm; and (4) the empire of oil has not done away with nature but unloosed new scientific and political desires for the natural world. Drawing these fields of inquiry and insight together, this book displaces the reason of the commodity as the methodological and conceptual basis for taming the deleterious properties of the combustible present. This book approaches the negative ecologies of the oil industry not as accidents condensed in time and space but as the fertile soil within which new political theologies of altered life take root. Gathering these concerns together, it advances a reappraisal of the environment for anthropological research and social theory, one that locates its critical capacities not as the best means of guarding against the rising disasters of oil but as a complicit project of those very disasters. What forms of research, critique, and mobilization, the book asks, may now be needed not only to more forcefully confront the deathly properties and looming foreclosures of fossil fuels but to envision ways of living beyond them? And how can we aspire to those new worlds while not losing sight of the injuries already sustained, the justice long overdue?

CHAPTER I

Environment

A Disastrous History
of the Hydrocarbon Present

Is the environment worth the effort? The environment often seems far too easy, far too obligatory, and far too footloose a concept to warrant serious attention. It somehow evokes both bookish abstraction and populist rousing, it cobbles together science and advocacy only to blunt their conjoined insights, and it continues to elude fixed definition even while basking in stately recognition. The banalities of this mess can give the impression that the environment has no real history, has no critical content, and heralds no true rupture of thought and practice. The environment, in the eyes of some, is mere advertising. If there is a story to the environment, others suggest, it's largely one of misplaced materialism, middle-class aesthetics, and first world problems. Such has been the sentiment, such has been the dismissal.

In the rush to move past the environment, few have attended to the history of the concept. This is curious, as the constitution of the environment remains a surprisingly recent achievement. In the late 1960s and early 1970s, the environment shifted from an erudite shorthand for the influence of context to the premier diagnostic of a troubling new world of induced precarity (whether called *Umwelt, l'environnement, medio ambiente, huanjing, mazingira,* or *lingkungan*).[1] While the resulting recognition of the environment largely consisted of bringing existing problems together under one umbrella—factory pollution, urban sewage, radioactive fallout, automobile emissions, garbage disposal, and even climate change—the resulting synthesis was powerful.[2] It was almost as if a light had been switched on to reveal a whole new world of toxic trespass. Such illumination—what historian Joachim Radkau (2014) has called "a New Enlightenment"—posed an unsettling provocation: perhaps progress was not achieved autonomy from the natural world but waves of profit and power undermining the very basis of life. As shorthand for the resulting crisis of life, the environment became an insurgent field devoted to understanding damaged life and taking responsibility for it. Despite scholarly attempts to bury the term within more established histories, the environment signaled something profoundly new for outraged citizens, concerned scientists, and savvy politicians.[3]

The novelty of the environment did not pass without notice. Celebrating its recent arrival, *Time* magazine named the "environment" the issue of the year in 1971. When fifty paperbacks on the discovery of the environment were published in 1970 alone, a *New York Times* reviewer described being "inundated by books on the environment" (Shepard 1970: 3). While drawing very different conclusions, these books almost uniformly noted how the environment drew the ailing "life support systems" of Earth into unsettling focus. Privileging the pragmatics of human survival over inherited precepts, the environment introduced an impending future as the new guidebook for moral conduct and political action in the present. A French minister called *l'environnement* an "imperialist" term for how quickly it infiltrated the country, demanding new oversight within the most ordinary of places and practices (Poujade 1975: 27). The family automobile, dishwashing detergents, and plastic bags found themselves suddenly shot through with planetary significance. Astounded at the range of the concept, ecologist Paul Shepard (1970: 3) insisted the environment "is genuinely new in its planetary perspective and connection to war, poverty, and social injustice." The novelist Isaac Asimov (1970: A9) summarized the sentiment: "Environment has become a magical word," he wrote in 1970, drawing together the ordinary and the planetary, our present plight and our future ends.

During the 1970s, the environment rather suddenly became "a household word and a potent political force," as one White House report reflected (CEQ 1979: 5). Prompted by a somber announcement from the UN secretary general that "it is apparent that if current trends continue, the future of life on Earth could be endangered," the United Nations (UN) organized a conference on the human environment in 1972 (UN 1969: 10; Ward and Dubos 1972). Within two decades nearly every government had commissioned a new agency or ministry to protect the environment. In the United States, environmental studies was inaugurated on college campuses across the nation, and major newspapers added an "environment beat." Whole subfields in environmental law and environmental science sprang up almost overnight. The *environment*—a term "once so infrequent and now becoming so universal," as the director of the Nature Conservatory commented in 1973 (Nicholson 1970: 5)—soon came to monopolize popular and scientific understandings of damaged life and the state's obligation to it. The environment emerged as "a crisis concept," write historians Paul Warde, Libby Robin, and Sverker Sörlin (2018: 23), born in peacetime prosperity pricked by a dawning sense of doom. Visualizing the synthetic webs underlying and undermining the modern project, the environment advanced a theory of entangled life beyond the nature/culture dualism. Vividly documenting the basis of life slipping just beyond the fixtures of modernist control, the environment offered a precocious theory of the Anthropocene. To the great misfortune of contemporary scholars scrambling for the title of first author, this early theory of manufactured disarray was most substantially advanced by the state.

While much as been made about how this crisis of life helped lay the affective groundwork for the rise of environmentalism (Gottlieb 1993; Worster 1994; Sellers 2012), much less attention has focused on the underlying materiality of this crisis.[4] As the resulting social movement holds the attention of scholars and citizens alike, the physical ruptures these campaigns responded to has drifted out of focus. Many of the specific problems that provoked what became known as "the environmental crisis" had their basis in what Rachel Carson (1962: 11) once called "the wonder world" of hydrocarbon innovation in post–World War II America. As two leading public health officials noted in 1955: "The recent advent of the atomic age, the era of synthetics, and the petroleum economy, when combined with epidemiological observations, indicate that a general population hazard is of more than theoretical significance" (Kotin and Hueper 1955: 331). By the 1960s, ecologists were learning to see just how thoroughly two icons of modern power—fossil fuels and the atomic bomb—had infiltrated the very fabric of life. Christened "our synthetic environment" by Murray Bookchin (1962), this scientific recognition of porosity and precarity punctured the modernist swagger of modular control. While the specific instances of injury were incredibly wide-ranging, the cause was surprisingly uniform: hydrocarbon emissions, petrochemical runoff, and radioactive fallout. In other words, the material basis of American prosperity and power in the twentieth century.

Resituating the environment around American ascendance places the emergent crisis of life and resulting structure of feeling on a more imperial foundation of disruption.[5] Rather than starting with the instigated social movement—environmentalism—and grasping the world from within its mobilized outlook or starting within the resulting domain of objectivity—environmental science and policy—and grasping the world from their already normalized overlay of toxicity and life, this book begins with the surplus of synthetic force that sparked the founding crisis. Emphasizing the messy materiality of "the environmental crisis" of the late 1960s can situate the protests and institutions that gave rise to the environment in the 1970s without too tightly bounding scholarly inquiry to their "post-material" suburban and state forms (Inglehart 1981). However provisionally, this also brings American empire into focus as an early provocateur of what Jane Bennett (2009) calls "vital matter" and others have taken up as the earth-shattering insight of nonhuman agency.

Many contemporary scholars newly smitten with the agency of the material world are swept up in a kind of utopian outlook, where a profound pessimism of the political conspires with a newfound optimism of the physical. Such work collapses all frustration with the shortcomings of existing politics into the fogged vision of human exceptionalism, suggesting that if we can only recognize the vibrant liveliness of the worlds beyond the human, a truly emancipatory politics will bloom organically from the rubble of the modern episteme. The most radical task, then, is to simply understand the world differently, to bracket a few centuries of history as comprehensive failure and look forward to the worlds to come. Perhaps this may hold promise with the ontological force of mushrooms, rivers, forests, and mountains, to name a few of the more consequential reworkings of materiality in anthropology today. But the sweeping optimism of this current of thought often ignores the more destructive agencies that enliven our present (or worse, may find misplaced optimism in their destructive capacities). What of the ontological force of toxic destruction and pandemic disease? Is the celebration of their agency also emancipatory? Or, paraphrasing Taussig (2018: 18), what if it is the viral terror of the contemporary that has endowed the natural world with a vitality that scholars only grasp in proliferating agencies divorced from history? The explosive force and "slow violence" of fossil fuels and nuclear weapons bring a very political history to these questions of agency, one saturated with the petrochemical and radioactive foundation of American empire in the twentieth century (Nixon 2011; Immerwahr 2019). Yet today, this history of synthetic force in the

FIGURE 2. 2014 Peoples Climate March. Photo by author.

imperial rise and reach of the United States, writes Adam Hanieh (2021: 28), "sits elephant-like within the ecological crisis of the present."

Such critical connections were closer to the surface during the rise of the environmental crisis. For writers like Rachel Carson and Barry Commoner, America's rising reliance on fossil fuels and nuclear weapons introduced a profoundly destructive agency into the world. This haunting corrosion of life came into analytical focus precisely for the open-ended harm it caused, and by foregrounding the entangled webs of harm unleashed by the fallout of petrochemicals and radioactivity, their public scholarship sketched out a still unfinished critique of American empire. In sharp contrast to the new materialism of contemporary scholarship, their writing disallowed any deferment of history in the reckoning with the futures already at work within us and refused any celebration of an emancipatory politics from the mere recognition of material agency. Moreover, the legacy of their work also demonstrates how a radical opening to material agency and entangled life did not, in itself, conjure a revolutionary politics so much as authorize new scientific and regulatory fields of technocratic control within the nation-state.

THE NEGATIVE ECOLOGIES OF POWER

In so many ways, the sudden and widespread realization of the environment was underwritten by the excessive synthetic materiality of American power. As many have argued, the properties of fossil fuels and atomic energy introduced a new material basis for coercive accumulation and authority and a new infrastructure for imperial projections of structural retribution and cultural aspiration that helped align the world with designs for a new American order.[6] Yet the unique properties of fossil fuels and atomic energy reached far beyond the coffers of corporations and the clenched fist of the state. Something of their force exceeded capture within positive iterations of wealth and influence. Something of their force made its way beyond capitalism and state power and into the fabric of life itself. And as they defied existing jurisdictions and disciplines and came to suggest worlds of consequence in gross surplus of their cause, these problems came to demand a new accounting.

Centering negative excess has a pedigreed intellectual genealogy. In 1947, Max Horkheimer and Theodor Adorno posed new questions about the excessive underside of modern power. Whether in trench warfare, administrated genocide, or suburban ease, an unprecedented union of "machines, chemicals, and organizational powers" ([1947] 2002: 184) promised to launch human might beyond the gravity of history and nature. For Horkheimer and Adorno, the concerted effort inaugurated a new "epoch of the earth's history" (184) founded on the divorce of jagged historical realities from the scientific pursuit of unfettered power, the privileging of a life of ease over any obligation to care, and the repression of an imploding present under the banner of a more perfect future. Reason metastasized from a ladder of critical enlightenment into the author of oblivious annihilation. Almost as if self-driven, the resulting "motorized history" (xv) rushed ahead of any political accountability, let alone revolutionary resistance. By way of the automobile, DDT, and atomic weapons, Horkheimer and Adorno sketched out "the calamity which reason alone cannot avert" (187). Instead, they wrote of the postwar moment: "The hope for better conditions, insofar as it is not an illusion, is founded less on the assurance that those conditions are guaranteed, sustainable, and final than on a lack of respect for what is so firmly ensconced amid general suffering" (186). And it is from the dark alleyways and deformed lives of the catastrophic ascendance of instrumental reason that such promising disrespect resides. (As Adorno wrote at the time, "automobile junk yards and drowned cats, all of these apocryphal realms on the edges of civilization move suddenly into the center," quoted in Buck-Morss 1977: 189). Such work resituates the negative not as a fundamental lack but as a provisional grasp of reality in defiance of tyrannical rationality.

Twenty years on, and Theodor Adorno only deepened his conviction of negativity as the most philosophically astute, politically uncompromised, and empirically potent realm of the contemporary. "After the catastrophes that have happened, and in view of the catastrophes to come, it would be cynical to say that a plan for a better world is manifested in history and unites it," wrote Adorno in *Negative Dialectics* ([1966] 2007: 320). Recoiling at the human fodder readily fed into the crackpot utopias of both liberalism and state socialism did not mean admitting critical theory could not counter the crisis at scale. It meant rooting critique in the historical necessity of change inoculated from the corporate dogma of technical progress. If no overarching spirit of redemption unified recent historical experience, there was still the possibility of a more encompassing home for critical theory. "No universal history leads from savagery to humanism, but there is one leading from the slingshot to the megaton bomb"(Adorno 2007: 320; see also Vázquez-Arroyo 2008; Chakrabarty 2009). Whether in lingering memories of concentration camps or rising fears of impending nuclear war (or, we might add, all of those places where the Cold War was anything but), a jilted sensibility of unbound destruction had become ordinary. "Absolute negativity is in plain sight" (Adorno 2007: 362). Such negativity may be ubiquitous, but such realities were also conceptually invisible. Negative realities were obscured by the omnipotent optimism of instrumental reason projecting redemptive futures. Yet for Adorno, the negative does not exist within the shadow of the positive. The rippling losses are not secondary, peripheral, subordinate, or ancillary to speculative growth. The losses were fast becoming the more definitive reality. How might centering negativity help prick the positivist plague of instrumental reason and stage a more potent politics of transformation? Not only is nonsensical destruction the direct consequence of the teleological pursuit of capital, but the injuries sustained grossly exceed the amassed gain. Unloosed ruin is the very landscape upon which categorical progress gains viral force and, on pain of waking up, cannot acknowledge. With revolutionary demurral, Adorno looked to negativity as proliferating instances tripping up the imperial conceit of progress. Against the omnivorous ideological appetite and relentless synthetic acceleration of capitalized history, a modern whirlwind that seemed capable of bending any inkling of conceptual optimism into widening projects of dispossession and devastation, the negative alone refused easy recruitment. Standing just outside the infectious mantra of mutually assured redemption through unrestricted force, the negative rebuffed philosophical appeasement and political reconciliation. It glimpsed an entirely different reality. Negative dialectics, for Adorno, roots inquiry and intervention within "the extremity that eludes the concept," for anything else would merely provide "musical accompaniment" to annihilation (365).[7] Far from a political identification with the forces of destruction or celebration of the baptism of coming catastrophe, laying the emphasis on negativity gave credence to the present impossibility of current arrangements. To negate the negative you first have to take it seriously. Prioritizing the bruised and broken came to seed grand delusions of fueled progress with "epitomes of discontinuity" (320) that tripped up the otherwise relentless acceleration of history. Inhabiting experiences of negativity offered critical theory a way to grasp the present from within its historical momentum but not determined by it: experiences of negativity repulsed incorporation into instrumental reason, anchored the apprehension of reality in common destruction without validating utopian redemption, and opened a doorway to seize upon radical possibility from within a motorized history already derailed. In this insistence on the priority of the negative, Adorno was not alone.

Eugene Odum's classic textbook, *Fundamentals of Ecology* (1953), oriented the emerging science of ecology toward the relations that energized life. Documenting a cycling between animate and inanimate, ecology found life suspended within intricately balanced circulations of matter.[8] A decade later, a number of scholars and writers turned Odum's insights away from the isolated mountain ponds that populated his textbook and toward the manufactured present. Rachel Carson, Barry Commoner, and others drew on ecology to query vital biochemical cycles being thrown into disarray by the noxious excess of factories, power plants, automobiles, fertilizers, and nuclear testing. Observing the ease with which petrochemical pollution and radioactive waste joined the chemistry animating life, often to disastrous effect, they centered their inquiry on "the webs of life—or death—that scientists know as ecology," as Rachel Carson (1962: 189) put it. This was an ecology that refused an outside to the industrial and militarized landscapes of the contemporary world. The subject of this renegade ecology was resolutely present tense: the split personality of American postwar society, possessed of both synthetic prosperity and biological precarity for which there was little precedent. Turning toward the historical present, ecology became "a subversive science," as Paul Shepard put it in 1969.

If the rise of ecology brought new emphasis on the biochemical relations that condition life (McIntosh 1987; Golley 1993) and cultivate a sum flourishing in excess of its parts, Rachel Carson, Barry Commoner, and others showed how ecology could move just as easily in the opposite direction: the pulsing relationality of life could also be its undoing. Toxicity generated subtractions well in excess of the component parts. This synthetic hijacking of what Georges Bataille ([1967] 1991: 27, 10) once called the "chemical operations of life" bent energetic excess away from "the effervescence of life" and toward its caustic dissolution. Here, a provisional commensurability took hold in worlds sundered from the modern teleology of progress and unified by their scientific proximity to ecological collapse. Theirs was a negative ecology, bearing witness to how rising waves of hydrocarbon emissions, petrochemical runoff, and radioactive fallout spilled into the fabric of life, contorting vital relationality into new intimate and planetary infections of injury and destruction. Reviewing the destruction underway in the name of progress, Rachel Carson (1962: 99) wrote, "The question is whether any civilization can wage relentless war on life without destroying itself, and without losing the right to be called civilized."

With reference to petrochemicals and atomic bombs, in 1966 Adorno sketched out the emergence of negativity within the philosophical history of Western capitalism. As *Negative Dialectics* took shape, Rachel Carson and Barry Commoner documented a similar plot within the emerging ecological crisis of American prosperity. "We are living in a false prosperity," Commoner (cited in Egan 2007: 141) told American audiences in 1970. "Our burgeoning industry and agriculture has produced lots of food, many cars, huge amounts of power, and fancy new chemical substitutes. But for all these goods we have been paying a hidden price." That price, Commoner concluded, was the systematic corrosion of life itself. For Carson and Commoner, the generalized toxicity of DDT and radioactive fallout had come to outweigh whatever technical domination they heralded in the specific eradication of insects or enemies. While Adorno turned to negativity at the commanding heights of theoretical prowess, it inspired few protests and even fewer policies. In *Negative Dialectics* Adorno carefully leaves room for negativity to become "a creative force in itself," as Susan Buck-Morss (1977: 36) has written, but Adorno himself did very little to either cultivate or delineate the insurgent possibilities of the negative (see also Gordillo 2014).

Carson and Commoner brought the negative into view in a way that inflamed popular enthusiasm for systematic change. Both Carson and Commoner privileged the sprawling webs of destruction in the wake of synthetic might as a way of defrocking the modern conceit that nature and history could be easily shed. And in so doing, they opened new fields of science, policy, and protest premised on first acknowledging the unbearable weight of multispecies suffering. If emphasizing negativity offered a philosophical counterpoint to the tyranny of capitalized rationality, it also offered an ecological counterweight to the American fantasy of petrochemical prosperity and thermonuclear stability.

Yet if Adorno was equally suspicious of both revolution and science, the work of Carson and Commoner would provide new cause to join their forces. Against the complicit sciences of extraction and extermination, a revolutionary science of care was born. Rejecting the "little tranquilizing pills of half-truth" found in the petrochemical industries patronizing claims to unrivaled scientific authority, Rachel Carson (1962: 13) insisted on a science attentive to mending the broken world. In accepting the National Book Award, Carson said the notion "that science is something that belongs in a separate compartment of its own, apart from everyday life, is one that I should like to challenge. We live in a scientific age; yet we assume that knowledge is the prerogative of only a small number of human beings, isolated and priestlike in their laboratories. This is not true. The materials of science are the materials of life itself. Science is part of the reality of living" (cited in Gottlieb 1993: 122). This science, what Barry Commoner (1967) later called the "science of survival," came to privilege the empirical weight of destruction over the imperial banner of progress, and also came to prioritize urgent care for the world over any discussion of institutional capacity or costs disproportionate to profit. Such a science was revolutionary not for its active use of the term but because in centering the negative excess of power, such work gave rigorous scientific definition to an emerging crisis whose only solution would be the radical transformation of society itself. "Our system of productivity is at the heart of the environmental problem," wrote Barry Commoner (cited in Egan 2007: 141). And overturning that system of productivity would be essential to any real solution.

The massive industrial, militarized, and then consumer expenditure of energy in twentieth-century America—an expenditure for which there is no precedent in human history (McNeil and Engelke 2016)—unloosed a synthetic excess that frayed the fabric of life. These expenditures, as Donald Worster (1994: 340) put it, quickly called into question "the elemental survival of living things," a profoundly new personal and planetary fact. As hydrocarbon and nuclear excess infiltrated interwoven existence, Rachel Carson, Barry Commoner, and other ecologists helped solidify an emergent sense of the vital interconnectivity of life and the recent achievement of its precarity. The resulting vision not only brought the ends of life into disconcerting focus, it also opened the routes, accruals, and cascading effects of those very ends into new scientific visibility and political responsibility. The resulting ecological accounting of power learned to see how elemental cycles that foster life were being infiltrated and impaired by synthetic force. It was a realization that brought empirical clarity to the historical contingencies of life on Earth and their groundbreaking possibilities once properly recognized. And soon enough a new field of science, a new jurisdiction of law, and a new social movement sprang up in this radical realization and what it demanded of the contemporary.

For many writers, scientists, and activists working in the 1960s, this dawning contradiction of life was brought into crisp focus by two compounds: DDT and strontium 90.[9] As the "most striking petrochemical technology success story of the postwar era," DDT was used extensively as an insecticide across towns and fields in the United States in the 1940s and 1950s and in many tropical regions (Gottlieb 1993: 82). In the 1950s, enough DDT was sprayed in America "to give every man, woman, and child in the country their own one-pound bag" (Dunlap 2008: 5). Despite targeted uses in the extermination of pests, DDT readily accumulated in living tissues of a wide variety of animals, including humans, "with the result that many unforeseen, irrevocable, and undesirable side effects have arisen on a sizeable scale," noted a report from the first American Medical Association congress on environmental health (White 1964: 729) A 1964 survey in the United States revealed that the "storage of DDT-derived material in body fat averaged 12.9 ppm for the general population" (White 1964: 729–30). DDT, an insecticide engineered from chlorinated hydrocarbons, is mistaken by the digestive tract of mammals for an ingredient of body fat and stored in fatty tissue. Unlike organic body fat, however, chlorinated hydrocarbons resist being broken down by internal enzymes and used as energy by the body. Once stored, they can became a permanent (and often perverse) part of the body. Bioaccumulation of DDT can contort reproductive functions, leading to high rates of miscarriages and birth defects. By the 1960s, DDT was showing up in most forms of life on Earth, including polar bears and penguins.

Conjured in the alchemy of nuclear detonation, strontium 90 dusted entire hemispheres (Rudd 1964). As "a chemical relative of calcium, it takes a similar biological course" (Commoner 1967: 15), moving from fields to dairy cows to human bones and teeth with startling rapidity.[10] Moving through the "soil-plant-milk-human chain," strontium 90 accumulated at alarming levels in the fast-growing bones of breast-feeding infants and developing fetuses, a process attributed to the rising incidents of leukemia in US children during the 1960s and 1970s (Kulp, Schulert, and Eckelmann 1957: 1249; Newcombe 1957; Eckelmann, Kulp, and Schulert 1958; Kulp, Schulert, and Hodges 1959; Kulp, Schulert, and Hodges 1960; Kulp and Schulert 1962). These discovered linkages led to the widespread collection and study of baby teeth as a crucial indicator of the spread and density of radioactive fallout in human bodies (Reiss 1961). Once unloosed upon the world, DDT and strontium 90 leaked out of projects of engineered control and into the operating system of both the planet and its inhabitants.[11] In their novelty and then ubiquity, in their promise and then perversions, DDT and strontium 90 inaugurated a recognition of the very fabric of planetary unity and, at the same time, the manufactured power that actively threatened it. The negative excess of DDT and strontium-90 turned the invincible reason of synthetic power into its opposite while foregrounding the urgency of planetary care.

As wholly fabricated substances, DDT and strontium 90 were "tracers," as ecologist George Woodwell (1967: 24) put it, that cast new empirical light on planetary systems and their interplay with food chains and cellular metabolism. Following DDT and strontium 90 gave new proof of the earthly entanglements of air, water, and soil that animated life. As Kate Brown (2013) and Laura Martin (2022) have shown in very different settings, following the routes of radioactive fallout offered definitive evidence of what had previously been the theoretical setting of ecology: ecosystems as deeply interwoven bio-chemical exchanges bound in place. This fallout also showed such earthly entanglements were never innocent of history. As a UN report later summarized, the observed pathways of DDT and strontium-90 "provided a solid basis for a completely new appreciation of the unity, interdependence, and precariousness of the human condition" (Ward and Dubos 1972). This novel empirical window of negative entanglement cracked open the possibility of an insurgent new science of the contemporary world. As DDT and strontium 90 helped revolutionize planetary and cellular optics, those same analytical advances bore witness to a subtle if systematic unraveling of life. Many of the resulting observational infrastructures—newly calibrated to measure and monitor the biochemical composition of planetary systems—laid the groundwork for understanding climate change (Edwards 2010; Masco 2010; Lepore 2017; Collier and Lakoff 2021). DDT and strontium 90 traced out a radically new understanding of a shared biosphere only to simultaneously show exactly how it was coming undone. Such an impact, at once historical through and through and yet just outside the registers of historical reason, wounded life not by frontal assault but by infiltrating the biochemical nexus that grounded life's possibility. This entangled reality trips up the reifications of the autonomous self and revered distinctions of nature and society that charter modernist projects of accumulation and authority. DDT and strontium 90 identified the filaments that stitched together spraying fields and fatty tissue, nuclear detonation and baby teeth, modern power and cancerous bodies. Such linkages were always more than a network revealed, for they also described a mutinous harm now freed of the existing script of near and far, subject and object, meaning and matter. As Rachel Carson (1962: 189) reflected, "There is also an ecology of the world within our bodies."

"The fallout problem," as Commoner (1958) so aptly named it, demonstrated how easily hydrocarbon and nuclear excess traversed disciplinary strictures, militarized borders, and species hierarchies to achieve near universal contamination (see also Masco 2021).[12] Such fallout also provoked an unsettling recognition: the force of fossil fuels and atomic energy is not fully expended at the moment of use nor wholly annexed into national wealth or military might. After the explosion, disruptions rippled across the fabric of life, making terms like *commodity* and *weapon* but flickering events in much more extensive landscapes of dissonance.[13] In their roving mobility, geological persistence, and affinity for cellular systems, the negative ecologies of DDT and strontium-90 overwhelmed modernist fantasies of modular control that directed their use. Not only did this suggest worlds of consequence far in excess of their founding form, but such dissonance provincializes materialist critiques still wed to the labored dimensions of the commodity, still anchored to the physical immediacy of violence.[14]

Nor was human life exempt from these forces. As Barry Commoner told students on the first Earth Day in 1970: "You are the first generation in the history of man to carry Strontium-90 in your bones and DDT in your fat: your bodies will record in time the full effects of environmental destruction of mankind." Defying any easy separation of nature and culture, these insights showed how easily hydrocarbon pollution and radioactivity could move through earthly mediums to injure distant plants, animals, and humans. As low-level exposure to petrochemicals and radioactive fallout was linked to cascading species decline and sharp upticks in cancer rates, such contamination came to prefigure a catastrophic universalism.[15] "Toxicity to humans is but one aspect of the pollution problem," ecologist George Woodwell wrote in 1970, "the other being a threat to the maintenance of a biosphere suitable for life as we know it" (431). In planetary fact if not yet in political practice, the negative surpassed the positive.

Although rarely framed in such a manner, growing concern with these "agents of death," as Rachel Carson (1962) described DDT and strontium 90, also advanced a recusant theory of materiality. DDT and strontium 90 commandeered the interdependence of life.[16] Unlike more recent reformations of materiality authored around cyborgs, microbes, mushrooms, rivers, and Indigenous cosmologies, the physical force of DDT and strontium 90 was not so easily allied with or channeled into projects of gain.[17] Theirs was a negative agency, one that by interrupting the conditions of life illuminated the contingencies of life, one born of power but never contained by it, and one whose near universal reach revoked any return to purity. Although rarely stated concisely, this emerging grasp of negative materiality also turned the relation of nature and culture into a historical dialectic whose impending synthesis would either be revolution or the end of life.

IMPERIAL FALLOUT

These negative ecologies—or "chains of evil," as Rachel Carson (1962: 2) named them—brought rising petroleum prosperity and thermonuclear statecraft into focus not as the pinnacles of American supremacy but as its very opposite: an unchecked regime of degenerative life. Fossil fuels and atomic bombs poured the concrete foundation of American swagger in the twentieth century. If the physical properties of sugar, cotton, and other cultivated goods helped shape previous iterations of empire (Mintz 1985; Beckert 2015), the twentieth century witnessed US ascendance taking shape in accordance with the physical properties of fossil fuels and atomic weapons. During World War II, extensive deposits of uranium in the American West and abundant reserves crude oil in California, Oklahoma, and Texas helped catapult the prowess of the US military to new planetary scales of mechanized violence. "Oil is ammunition," ran one government poster during the war (cited in Huber 2013: 71). Petroleum gave the US military unprecedented mobility in

the oceans and skies of every continent, while nuclear weapons provided unprecedented powers of regional annihilation at the flip of a
switch. After World War II drew down, concerted efforts to keep the
profitable spigot flowing engineered that same surfeit of crude oil into a
new American infatuation with big cars, suburban ease, and cheap food,
while a surplus of nuclear weapons encased rising domestic prosperity
within the promise of invincible security.

By fate of geography, North America happened to be home to some
of the richest deposits of crude oil in the world. In 1945, "two of
every three barrels of oil was produced in the United States" (Moore
and Patel 2018: 176), and up until 1970 oil production in the United
States dwarfed that of every other nation. In contrast to almost every
other major oil-producing nation, this American bonanza was rarely
exported after the war (EIA 2021). Instead, its excess was channeled
into every facet of American life, where the relative cheapness of energy
propped up an infectious feeling of momentous progress and birthed
a new benchmark for the good life: a house for every family, car in
every driveway, and meat on every plate. As Matthew Huber (2013:
56) documented, in the 1950s American policies, advertising, and common sense all "*equated* petroleum consumption with a high standard of
living." In 1965, the United States consumed a whopping third of the
world's energy despite comprising only about 5 percent of the world's
population (McNeill and Engelke 2016: 10). After World War II, US
per capita energy expenditure rose to nearly seven times the world
average, and twice as much as comparable European nations (Rosenbaum 1977). This abundance of fossil fuels was built into the design
and operation of the American suburbs—"the greatest misallocation of
resources in world history," according to James Kunstler (2005: 233)—
and its condition of possibility: mass ownership of the automobile
(Mumford 1963; Reisman 1964; Jackson 1985; Wells 2012). In 1972,
half of the largest corporations in America derived the bulk of their
revenues from this oil-automobile-suburbs complex, while employing a
significant portion of the American working population (Sweezy 1972).
The American automobile and suburban home were both incredibly
energy inefficient compared with European and Japanese counterparts.
Yet inexpensive petroleum and plastics allowed both to provide oversized ways of living for the masses, helping shift the political priorities
of labor from claiming the collective power of production to defending private palaces of consumption (Cohen 2004; Huber 2013). A deluge of cheap energy helped catapult a new aristocracy of labor into a

splendor previously only available through imperial theft of resources from elsewhere (DuBois 1915; Lenin 1920; see also Bond 2021c). While this fueled complex of high sector wages and petroleum cheapened food and housing provided uplift for many white, semiskilled workers, most Black and agricultural workers were barred from participation in this charged American Dream (Davis 1986). Suburbs may be "bourgeois utopias," as Robert Fisherman (1987) once quipped, but they are utopias that only came to feel within reach for the white working class through prodigious expenditures of energy.

While American housing and transportation were privatized through the glut of petroleum, the pliability of petrochemicals instigated a new autonomy from natural resources (Schnaiberg 1980: 120). Through the alchemy of petrochemicals, synthetics came to replace rubber, quinine, cotton, and a host of other tropical resources built into modern progress as "the laboratory replaced the land as the source of materials" (Immerwahr 2019: 274; see also Hanieh 2021). "A New World," effused a Mobilgas Ad, "is Being Born in America's Petroleum Laboratories!"(cited in Sheller 2019: 66). Or, as Shell Oil executives championed, "Plenitude from Petroleum"(cited in Huber 2013: 90). If petrochemicals lessened the need for colonial plantations in the tropics, they intensified the cultivation of domestic land through supercharging agriculture with industrial logics and chemical coercion. Turning fossil fuels into cheap food happened first through the mechanization of cultivation (Fitzgerald 2003) and then through hybridized crops that could forgo human care with generous application of petrochemical fertilizers and pesticides (Kloppenburg 1988). Reviewing this history, Matthew Huber (2013: 87) concludes, "The American food system has been completely fossilized." In this trial run of the Green Revolution, Jason Moore (2015: 251) has pointed out, labor inputs in American agriculture "fell by more than two-thirds" between 1935 and 1970, while "fertilizer and pesticide inputs increased by an extraordinary 1,338 percent." Massively inefficient in terms of energy use but obscenely profitable in terms of capital returns, these dynamics helped push nearly four million farms into insolvency while consolidating control of agriculture into a handful of corporations. Pointing to the "petrochemical-hybrid complex" at the heart of industrial agriculture in America (and at the heart of the developmental model the US exported to the world on pain of financial ruin in the Green Revolution), Moore (2015: 251) and others identify the capitalized synergies between petrochemicals and cheap food as the launchpad of American influence in the twentieth century.[18] Whether through the profitable inefficiencies of eight-cylinder cars, suburban detachment, or carnivorous meals, the American Dream was brought to life in the secular catechism of fossil fuels as abundant, cheap, and inconsequential.

Securing this American effervescence was the muscular threat of excessive nuclear expenditures elsewhere. In the two decades following World War II, the United States both exponentially increased its stockpile of nuclear weapons—from 300 in 1950 to nearly 20,000 in 1960—while controlling roughly 90 percent of the world's nuclear weapons (Norris and Kristensen 2010). This amassed atomic firepower grossly exceeded any tactical military purpose. In a recently declassified report from the dawn of the arms race, scientists at Los Alamos estimated "it would require only in the neighborhood of 10 to 100" concerted thermonuclear detonations to wipe out the human species (quoted in Bienaimé 2016: 1). By 1975, the United States had 27,500 nuclear weapons. For most of the Cold War, the United States had enough nuclear firepower constantly at the ready for instant launch from air, sea, and land—and encased in an archaic system of automated and irreversible authorizations prone to error (Schlosser 2013)—to annihilate a dozen planet Earths. Flexing this weight of unimaginable devastation helped persuade a new form of global compliance. Whether in reserves of crude oil or arsenals of nuclear weapons, such abundance of synthetic force inaugurated a lifestyle of consumer bliss backed by world-ending violence.

This American infatuation with synthetic force advanced a new imperial methodology for resources without colonies and coercion without occupation.[19] So long as ample supplies of uranium and petroleum were secured, the United States "replaced colonies with chemistry" as the primary engine of American ascendance, as Daniel Immerwahr (2019: 271) argues. At the same time, a global network of military bases supported flying fortresses and submarines that threatened catastrophic violence everywhere. Fossil fuels and nuclear weapons appeared to emancipate projects of accumulation from any reckoning with earthly matters. American empire hit its stride on the world stage dripping with the harnessed might of petrochemicals and radioactivity.

Thermonuclear statecraft and petro-capitalism poured the material foundation of US empire in the twentieth century. Yet even as they underwrote the global conceit of American power, each mapped America's imperial structure in strikingly different ways. Each took momentous shape toward a divergent purpose, with distinct fields of operation, conceptual architectures, and amassed influence. The vast expenditures required to manufacture, store, and launch the atomic bomb are hard to explain within accounts that take capital as the only game in town. The atomic bomb requires some acknowledgment of the autonomy of the state. On the flip side, fossil fuels are nearly impossible to explain without a view of capital and the transnational corporation.

Critical scholarship that describes petroleum prosperity as the engine of US empire looks to the primacy of capitalistic extraction, petrochemical bounties of cheap food imposed at home and abroad, critique blunted by suburban accumulation, and the transnational corporation as the ascendant iteration of American empire. Critical scholarship that describes thermonuclear statecraft as the engine of US empire turns to the relative autonomy of the state, political logics of protection and destruction, critique blunted by the affective cultivation of fear, and the patchwork of presences that place the US military within striking distance of anywhere and anyone as the ascendant iteration of American empire. Critical social research has often followed this bifurcation in critical explanations of US empire in the twentieth century. There may very well be analytically distinct explanations (not to mention opposed political priorities and theoretical implications) for emphasizing either the role of petro-capitalism or the role of thermonuclear statecraft in the rise of US empire. But there is striking commensurability in their ecological effects.

Whether from carbon heavy forms of suburban life, petrochemical infusions of cheap food, or radioactive shadows of flexed military prowess, the fallout of American power commingled in an emergent condition of degenerative life. Crude oil and nuclear weapons, in different ways, were enlisted into the American project for their positive accruals of profit and power. Yet the negative ecologies of that project soon exceeded existing measures of gain and infrastructures of control. Despite overwhelming release within American-backed efforts, DDT and strontium-90 were readily detected in nearly every population on Earth by 1970. Rachel Carson is one of the first to grasp this conjoined fallout as a new field of scientific inquiry, embedded ethics, and political engagement.[20] Many have followed her opening.[21] Privileging the overlapping negative ecologies of fossil fuels and nuclear weapons as a unified crisis moved toward a revolutionary science, for the resulting grasp of reality made the impossibility of the status quo as clear as a lit stick of dynamite.

This ecological grasp of the unified fields of petrochemical and nuclear fallout provided the charter jurisdiction of the environment. When leading academic journals proved reluctant to publish scientific reports on the proliferating instances of "fallout"—whether from pesticide runoff, automobile emissions, factory effluent, or nuclear blasts—Barry Commoner started his own journal to do just that. Its name? *Environment*. The environment, as scholars have noted, brought together the "conceptual association of various risks" (Radkau 2014: 100); in it "a whole cohort" of problems "were grouped together and labeled" environmental (Mahrane et al. 2012: 128). This has led some, like David Harvey (1996: 118), to bemoan the environment's "fundamental incoherence as a unitary concept." But in giving provisional thematic unity to the manifold examples of fallout coming into focus, perhaps such apparent incoherence was also its analytical and political strength. As Raymond Williams (1976) once quipped, words that matter most are those whose definitions we cannot agree upon, for in their definitional disputes they signal fights still unresolved in meaning and in matter. For these reasons, Williams famously named *nature* as the most complex word in the English language, and also perhaps the most potent. But as Christopher Sellers (2012: 9) notes, "since the midsixties, the term 'environment' has made a run on nature's crown" as most convoluted. Part of this certainly revolves around the unstable figure of fallout in providing the coordinates of the environment, but we should also note: these definitional questions emerge alongside the rise of twentieth-century US empire.

America's predilection to bend reality to its interests through generous applications of petrochemical and atomic force comes into view as an ecological debt already on its way to extinguishing human life on the planet. And it is in the disastrous effects of petrochemical prosperity and thermonuclear statecraft—what I call the negative ecologies of power—that the environment first takes empirical shape, first as the working title of the resulting crisis of life and then as a revolutionary science firing shots across the bow of reckless American materialism. Only later does the confrontational politics of negative ecology shift to the concessional politics so familiar in environmental science and policy today. A very worthy question remains as to why this rising environmental awareness—and its scientific attention to the material basis of American power—did not join more forcefully with decolonial movements against US empire worldwide. The point here is simply to point out the provisional commensurability of the critique.

Summarizing the implication of this empirical awakening to the ecological reach of power, Donald Worster (1994: 341) has written: "The only appropriate response was revolution." Tracing out the compounding impacts unloosed by the indiscriminate use of petrochemicals or nuclear weapons amassed overwhelming evidence of the scientific necessity of radical rupture. What is perhaps most surprising about this call to arms, Worster (22) notes, is that this insurgent movement was led not by artists or intellectuals but "by people within the scientific community." While perhaps not conversant in critical theories of revolution and moved more by facts than slogans, this "science of survival" nonetheless advanced an incisive grasp of the accelerating impossibility of the present and the absolute necessity of breaking with the synthetic pedestals of US power: fossil fuels and atomic weapons.[22]

In this, the negative ecology of Carson, Commoner, and others prompts a critical question: perhaps the world is becoming materially enlisted in the orbit of American power less by exploitation than by exposure. It is not always what is extracted that defines contemporary imperialism. James O'Connor (1998) has argued that fossil fuels are the greatest labor-saving device ever devised. As fossil fuels shift the wellspring of accumulation away from the exploitation of labor to the extraction of energy, they also shift the terrain of contradiction (O'Connor 1991; Foster 2009; Harvey 2014). The exhaustion of soil and the worker's body that define Marx's metabolic rift here become uncoupled from venues of exploitation: factories and fields. The existential coordinates of the metabolic rift become a more generalized condition (Foster 2000; Moore 2015), a condition that exceeds even the geography of capitalism (Brown 2013). The drift of pesticides, the heavy haze of smog, the scars left by extraction, and the hemispheric fallout of the bomb introduce sprawling new coordinates of the exhaustion of life. As Commoner described it, "the environmental crisis is an extension of the problems that were once confined to the workplace" (quoted in Egan 2007: 147). Whether in the rippling reach of fossil fuel combustion or nuclear explosions, exposure opens a new "scientific standpoint" from which to critique contemporary operations of power without either normalizing their structure or imagining an utopian outside to them (Lukács 1971). Centering the newfound vulnerability of life in theory and practice necessarily demanded a toppling of the structure causing it. Moreover, the radical standpoint enabled by the viral destruction of radioactivity and petrochemicals was broadly available in historically unprecedented ways. As the *New Left Review* commented on socialist debates over the nuclear bomb in 1982 (viii): "Planetary destruction affects all classes, as it does all societies. It poses the question of a common humanity *before* the advent of the classless society that socialist thought has always insisted could alone realize it."

The first Earth Day was a big tent organization, including both revolutionary and reformist orientations. Barry Commoner routinely showed how the environmental crisis might provide a "common ground" for all other progressive social movements in the late 1960s (Egan2007: 13). For it was in the unique ability of the environmental crisis to identify the material structure of US power that specific instances of injustice—"racial inequality, hunger, poverty, and war"—might be effectively linked and confronted together (Egan 2007: 118). Commoner argued that confronting the source of petrochemical and radioactive fallout would necessarily strike at the engine room of American empire. Moreover, Commoner saw the real crisis could be found not in population explosions but in the "civilizational explosion" found exporting the American way of life as the global benchmark of progress (Egan 2007: 125). If the Cold War taught Americans to be afraid in ways that amplified the authority of the security state (Masco 2008), Carson, Commoner, and other ecologists worked to introduce a fear that only the empowerment of people could solve.

In documenting the negative ecologies of petrochemical and atomic fallout, Carson, Commoner, and others cracked open a radical new understanding of the materialist basis of American imperialism and its ecological discontents. And in the long shadows cast by America's fueled ascent, negative ecology identified a chink in the armor. Of course, neither Carson nor Commoner nor those they inspired worked out the implications of negative ecology in this way. But their work gestures in this direction, and however tentatively, sketches out a new subversive science of American empire. If fossil fuels and nuclear weapons advanced new methods for manufacturing resources without colonies and exerting influence without occupation, they also introduced a new coordinate of racism in the everyday work of empire: unequal exposure. Such distributed harm might re-center the question of theft in the contemporary operations of empire.

As Timothy Mitchell (2011) has argued, the fueled autonomy from the commercial primacy of natural resources helped shift the primary domain of governance from colonial empires to national economies. Backed by coercive monetary exchanges underwritten by the cheap abundance of American crude oil and new global institutions safeguarding American interest, after World War II the United States came to champion decolonization (provided it didn't disrupt the continental legacy of settler colonialism and chattel slavery) and came to espouse a theory of the nation-state as the exclusive locus of political sovereignty and economic growth. Unlike accumulation premised on imperial conquest, the newly consecrated national economy "could expand without getting physically bigger" (Mitchell 2011: 139).

In this energized transformation of American power, however, perhaps the imperial coordinates of primitive accumulation did not disappear so much as they morphed from surface registers to subterranean and temporal ones. As Jason Moore (2015: 253) has argued, the twentieth-century rise of American dominance is marked by a "subterranean thrust" that coercively secures "prodigious volumes of cheap energy and cheap water." The geography of imperial accumulation shifts "from the primarily horizontal to the primarily vertical," Moore (2015: 254) writes, "not from one continent to another [. . .] but—primarily—from one geological layer to another." Crude oil is at the center of Moore's (2015: 252) account of this longer imperial history of the Green Revolution, as is the resulting "toxification" of the land. We might also add a temporal dimension to the physical footprint of American empire. With reference to nuclear winter and global warming, the energetic ascendance of the United States during the twentieth century may have avoided the colonial theft of land in the flexing of its imperial might, but it did so by stealing from the future.

Such theft is readily visible in the lives brutally shortened by exposure to uranium mining on Navajo lands in New Mexico (Pasternak 2011) or nuclear blasts in the South Pacific (Johnston and Barker 2008), to the petrochemical runoff of industrial farms in migrant communities in the Central Valley (Holmes 2013; Horton 2016), to the asphyxiating emissions in the Black neighborhoods lining refineries along Cancer Alley in Louisiana (Allen 2003; Singer 2011) or hazardous waste incinerators across America (Bullard 1990; Checker 2005; Ahmann 2018), to the leaky borders of nuclear weapons laboratories (Masco 2006; Brown 2013), to the corrosive shadow of the plastics industry (Altman 2022), to the toxic housing provided to victims of natural disasters (Shapiro 2015), to the catastrophic offshoring of the American petrochemical industry into more pliable places (Fortun 2001), to the generational violence of Agent Orange in Vietnam (Wilcox 2011), to the chemical defoliation tactics of drug enforcement in Latin America (Lyons 2020), to the exported dependence on petrochemical fertilizers and pesticides under the banner of a Green Revolution (Shiva 1989), to cheap packaged food and the surge of health issues in Belize and beyond (Moran-Thomas 2019), to the burning of electronic waste discarded by America (Little 2022), to the American-fueled and -armed Israeli occupation of Palestine (Weizman 2007; Stamatopoulou-Robbins 2019; Khayyat 2022), and to the blasted warzones in Iraq and Afghanistan now seeded with uranium-tipped shells and the toxic detritus of American war (Logan 2011; Jones 2014;

MacLeish and Wool 2018; Lutz and Mazzarino 2019; Rubaii 2020). Such pillaging of the future is now equally visible in the historical contortion of earth systems now tilting just beyond the conditions of multispecies flourishing and human dignity, whether in melting Arctic homelands (Krupnik and Jolly 2002; Watt-Cloutier 2015), ominous sea level rise (Marino 2015; Cons 2019), epidemics of extinction (van Dooren 2016; Parreñas 2018), ocean acidification (Kolbert 2006), untamable droughts (Bessire 2021), runaway firestorms (Petryna 2022), or a growing frequency of brutally destructive superstorms. Fossil fuels and nuclear weapons promised easy progress, yet such promises continue to fall on deaf ears within the lives and landscapes cut short by the negative ecologies of American power. In the most intimate and planetary of scales, the synthetic might of American imperialism in the twentieth century did not displace the physical footprint of empire so much as shift the coordinates of savage accumulation from the theft of space to the theft of time. The environmental crisis of the 1970s—with its unapologetic emphasis on the diminishing future as a new basis for contemporary ethics, science and politics—voices, however provisional, a critique of this newfound domain of imperialism.

AMERICA FIRST

The crisis of life brought into focus by the fallout of American power soon became a public event. As a genre and a mood, negative ecology resonated with a public that felt a newfound vulnerability amid relentless prosperity and progress, even if they did not yet have a firm grasp on its specific cause and consequence. Drawing attention to the conditions of life imperiled by fossil fuels, petrochemicals, and nuclear weapons, the work of Rachel Carson, Barry Commoner, George Woodwell, and others helped catapult "pollution" to the top of public concerns in opinion polls in the United States and Europe. Whether in sprawling cities and manufacturing hubs suffocating in smog, in plantations and farmlands saturated in pesticides that refused to stay put, in suburbs and slums doused in insecticides, or in swathes of entire continents and island homelands dusted with the seething remnants of nuclear detonation, this moment was marked by a grim awareness that the world was beset by emergent forms of generalized harm. As Ulrich Beck (1993: 72) has written of this rising apprehension, the disenchanted world of modernist control found itself newly enchanted by toxicity: "The role of spirits," Beck writes, is "taken over by invisible but omnipresent pollutants." For many Americans, oil spills became a potent image that conveyed a new reality of living in a prosperity overcome with its own crude waste. Disasters like the 1967 *Torrey Canyon* tanker spill or 1969 Santa Barbara blowout provided a potent visual for the rising awareness of living in a world haunted by synthetic toxicity. As two contemporary observers noted in 1973: "Oil slicks generally are more easily perceived than is the presence of toxic substances and visibility precipitates and intensifies public indignation" (Lleyellyn and Peiser 1973: 4). "Petroleum has become a devil in our civilization," effused one 1967 *New York Times* profile of a new kind of disaster: coastal oil spills: "Whether in a single dramatic incident or slowly, by default, it is fouling the seas, creating a survival issue both for sea life and for man himself" (Rienow and Rienow 1967: 25). Oil spills in Santa Barbara and elsewhere provided a searing aesthetic "fusion of fact and feeling, science and spectacle," writes historian Finis Dunaway (2015: 43–44), as national coverage "described blackened beaches and oil-covered wildlife as evidence of the escalating dangers of the environmental crisis." Oil spills distilled the image of life drowning in the muck of American excess. Prominently covered on front pages and broadcast into American living rooms, oil spills provided a salient dialectical image to the negative ecology of American power (Benjamin [1940] 2002; see also Taussig 2000).

If oil spills provided the popular aesthetic, ecology provided the analytical lexicon and affective register. As ecologists began tracing out the intersecting routes and intimate accruals of petro-prosperity and thermonuclear statecraft, they found a host of problems lurking just beyond the pale of cognitive genres, moral codes, legal strictures, and political institutions. In what Raymond Williams (1977) might call a "structure of feeling," an anxious mood rose up in the widening gap between the calm assurances of the state and the worsening state of the world.[23] Although here the negative ecology of Rachel Carson, Barry Commoner, and others surpassed the literary imagination to give a new language, experiential orientation, empirical texture, and moral outrage to this emergent haunting.[24] With fifty new paperbacks on the environmental crisis in 1970 and *Time* magazine awarding the "environment" the issue of the year in 1971, these issues found a broad audience. With varied arguments, evidence, and conclusions, this welter of popular attention to the environment brought the ailing "life support system" of Earth into stark, scientific focus. The feasibility of the future, displacing the precepts of the past, came to enthusiastically orient a wave of new moral conduct and political action in the present. Drawing what appeared to be dislocated injuries into wider networks of attribution, negative ecology advanced a new vocabulary of pathways of exposure (*fallout, bioaccumulation, web of life*) and zonal injuries (*excess deaths, sacrifice zones, ghost acres* and *dead zones*), while terms like *contamination* and *pollution* enlarged their meaning from policing the sexual boundaries of race to policing the biochemical boundaries of toxicity. While these fears reanimated fantasies of an untouched past, they also turned many toward present protests and questions of how to survive an impending planetary future. "The entire ecology of the planet is not arranged in national compartments," wrote George Kennan in 1970 (191–92). Pleading for partisan national, military, and corporate interests to stand down, Kennan (198) argued that we must place the survival of humanity—and our "animal and vegetable companions"—at the heart of the present crisis in order to privilege the necessary transformation. New publics cohered around these toxic uncertainties and fierce aspirations to change things, sometimes within a posture of solidarity as broad as the planet but just as frequently with one scaled to existing hierarchies of race, class, and citizenship.[25]

It is no surprise that many of these concerns first found voice in the United States. Perhaps no other society in human history has developed such a ravenous appetite for energy as the United States in the wake of World War II, when the relentless wartime expenditure of fossil fuels, petrochemicals, and atomic energy did not ratchet down after hostilities ceased but shifted weaponized force into suburban affluence. In postwar America, petroleum-saturated consumption and nuclear overtures of security became the distinctive American way of life in the twentieth century (Huber 2013; Masco 2015). Whether by reason of robust unions, embedded liberalism, or family values, many intellectuals and political platforms in the United States look back at postwar prosperity as the model of the good life that we should continually strive to recreate. For the Left and the Right, the 1950s often anchor the normative format of American political possibility. But perhaps the more material explanation of postwar prosperity lies in the unbroken American gluttony for crude oil and new development projects that dressed up unequal ecological exchange in the language of humanitarianism and structural adjustment (Hornberg 1998; Martinez-Alier 2002: 214), yet even this exchange was powered by the new fossil-fueled capacity of oceanic transportation and the new place of petrochemicals in export-oriented agriculture.

In the postwar era, the United States came to consume a mind-bogglingly massive amount of crude oil. Consumption grossly exceeded historical need as new expenditures of energy were dreamed up willy-nilly: new homes, new technologies, new diets, and new cities were built on the possibilities of this unending surplus: "Petrochemical America," as one apt review has called this new regime of life (Misrach and Orff 2014). As Henry Ford II quipped, "Minicars make miniprofits" (quoted in Egan 2007: 143).

In 1970, this turbocharged model of accumulation crashed into two unexpected roadblocks. Previously taken as boundless, the physical properties of crude oil screeched into view from two sides: an event of scarcity relative to soaring domestic demand and growing analytical recognition of the ecological fallout of unrestricted energy use. In 1970, domestic supply of crude oil in the United States peaked. Indifferent to such matters, petroleum-fueled lifestyles and social infrastructures continued unabated. The resulting scarcity took shape in the collision of flattening domestic supply and a relentless surge of heedless consumption. During the decade when American crude production began to falter and fall, overall American consumption of petroleum actually doubled. As others have argued, this was not a natural limit so much as a scarcity manufactured in the profitable inefficiencies of eight-cylinder cars, petroleum subsidized cheap meat, and the suburban atomization of consumption (Mitchell 2011; Jacobs 2016; Novy 2020). While the federal government ordered spigots of major oil fields opened wide, the stupendous American reserves of crude had passed their prime. As Bryon Tunnel, chairman of the Texas Railway Commission, commented at the time "Texas oil fields have been like a reliable old warrior that could rise to the task, when needed. That old warrior can't rise anymore" (quoted in Egan 2007: 151). The secretary of commerce put it another way: "Popeye is running out of spinach" (quoted in Egan 2007: 151). Discoveries in Prudhoe Bay, Alaska, and new exploration in the depths of the Gulf of Mexico promised to alleviate the crunch, but both would take years to come online, and neither could do more than slow the deficit. Something had to give.

At the same time, the cumulative impacts of a flippant reliance on petrochemicals, fossil fuels, and nuclear weapons lurched into broad public view. The percolating work of Rachel Carson, Barry Commoner, and so many others convincingly traced out a new world of consequence to the petro-prosperity and nuclear security of postwar America. This new mood was widely distilled in the spectacular imagery of oil spills (Morse 2012). The nation faced a dilemma: either recognize natural limits or compel oil from elsewhere (Jacobs 2016). The United States debated whether to redesign American life around alternative sources of energy, efficiencies achieved through federal investments in public housing and transportation, and drastically curtailed military expenditures of fuel, all of which were key platforms of the first Earth Day in 1970 (Gottlieb 1993; Rome 2013), or, in the other direction, to throw the weight of the federal government into the deregulation of the oil industry, military support for transnational oil companies, and a more imperial pursuit of foreign oil.

For many progressives, this proved the perfect storm to provide an enthusiastic mandate for the great transformation. The fossil fuel "energy system" was "cannibalizing" US society, as Barry Commoner described it in an interview with Studs Terkel (1979: 1). Reviewing the overwhelmingly ecological and economic case for breaking America's addition to fossil fuels and investing in a more sustainable future, Commoner thought a revolution was within reach. The sweeping ferment of Earth Day in 1970 demonstrated broad excitement for change, bringing together a startling array of constituents—white and Black, urban and rural, students and workers, young and old—under the banner of foregrounding the environmental crisis and transformation of American society it demanded (Gottlieb 1993; Rome 2013). "The energy crisis signals a great watershed in the history of human society," Barry Commoner commented (quoted in Egan 2007: 155).

Instead, President Richard Nixon opted for the salvation of foreign oil. As one White House adviser put it: "Conservation is not a Republican ethic" (quoted in Jacobs 2016: 43). Previously spurned, imports of crude oil doubled between 1967 and 1973 and came to provide "a safety valve" to defuse the reckoning that beckoned (Jacobs 2016: 39).[26] The subsequent Organization of the Petroleum Exporting Countries (OPEC) embargo of 1973–74 did not dissuade this energy policy so much as it enlisted the American military into it. It also realigned organized labor to the right of the environment. While Earth Day had garnered the support of many unions—who sensed the overlay of conditions inside the factory and conditions outside the factory—the OPEC embargo tilted many industrial unions toward a more hawkish support of foreign policy in the place of environmental commitments (Gottlieb 1993; Jacobs 2016). American dependence on foreign oil came to take on its own imperial geography of extraction as the per capita consumption of petroleum in the United States continued its unearthly rise. The global pursuit of more oil encouraged an even more deliberate overlay of American foreign policy and transnational oil corporations (Coll

2012). Not only was the American domestic addiction to petroleum now reliant on oil from abroad, but so were the American armed forces. Since the 1970s, the US military accounts for between "77 and 80 percent of *all* US government energy consumption" (Crawford 2019: 4). The American armed forces constitute the largest single institutional consumer of hydrocarbons in the world (Nuttall and Brazilian 2017). As General David Petraeus said in 2011, "Energy is the lifeblood of our warfighting capabilities"(quoted in Crawford 2019:1; see also Sheller 2019). Today, the US Army, Navy, and Air Force comprise "one of the largest climate polluters in history, consuming more liquid fuels and emitting more CO2e (carbon-dioxide equivalent) than most countries" (Belcher et al. 2019: 76). The planetary reach of the US military—essential to American efforts at a lighter touch of exerting influence without invasion—itself became dependent on more thuggish means of compelling cheap oil from elsewhere (Harvey 2003).

As Nixon opened the floodgates to cheap foreign oil, his administration also inaugurated the new responsibility of government to manage the emerging crisis of life (hoping, rather cynically, to deflect young voters' attention away from the debacle in Vietnam). As the contradiction between those two commitments of the state grew, the Nixon administration worked to defuse the more revolutionary potential of negative ecology by bending its insights away from assailing the material foundation of American empire and toward a more accommodating science providing modest guardrails for uninterrupted consumption: the environment we recognize today.[27] Reflecting on Barry Commoner's disappointment as the status quo marched on, biographer Michael Egan (2007: 155) writes, "American optimism was incapable of recognizing limits."

Whether in the suburban home, the ubiquitous automobile, the industrialized farm, the sprawling city, or the globetrotting military, the model of society in the United States was completely retrofitted, inhabited, and proudly held up as a universal model of prosperity on the trending assumption that hydrocarbon energy was cheap, copious, and inconsequential.[28] Since at least 1970, each of those assumptions has been persuasively dismantled. Yet consumption of fossil fuels in the United States continues skyward. How? As the Nixon administration intervened in the energy crisis of life during the 1970s, environmental science and policy came to separate the underlying addiction from the resulting impairment, effectively ignoring the underlying materiality of the problem. The disastrous properties of fossil fuels were externalized as autonomous fields of measurement and management.[29] These innovations privileged regulatory jurisdiction over ecological relations, engineered neutrality over material confrontations, and complicit facts over revolutionary science as they sought to erect technical barriers to the most egregious levels of pollution. Drawing together an older genealogy of thresholds with new pedagogies of technical planning, these new fields of science and law displaced the radical implication of the environmental crisis to objectify the problem entirely within a new administrative domain: the environment. These nationalized fields of science and law melded with surging public anxiety as the environment became not just an analytical operation but also a moral disposition aimed at ensuring prosperity while preventing the worst. If the environmental crisis of the 1960s insisted on resolving the contradictions between nature and culture, the new institutions of the environment in the 1970s instead found novel ways of stabilizing and managing their overlay.

Insisting that we only measure what was within the ability of power to mend, negative ecologies were domesticated into environmental science and policy. If negative ecologies brought scientific illumination and popular attention to a haunting materiality just beyond the ability of companies and states to resolve without undermining the basis of their own authority, the official recognition of the environment came to tether effective knowledge of these problems to pathways of action that left the pedestals of power wholly intact.[30] The environment shifted from a revolutionary reality striking at the heart of American power to an administrative domain to deepen the material dependence on fossil fuels without toppling the unsteady structure built atop it. The insurgent science of survival became an institutionalized science of concession. Much of this transition had to do with the consecration of two methods that together objectified the environment for defanged science and policy: thresholds and impact assessments.

THRESHOLDS OF TOXICITY

Peter Sloterdijk (2009: 18) claims that the environment came into being on April 22, 1915, in Northern France. He writes: "The discovery of the 'environment' took place in the trenches of World War I" with the advent of gas warfare. For it was when the basic conditions of human biology like breathing were turned into weapons that "the primary media for life [. . .] became an object of explicit consideration and monitoring."[31] Perhaps. But just as battlefield forces were learning to wage war with air and water, new government agencies were learning to regulate the toxicity of those same mediums inside the factory. While those on the battlefield sought to mobilize toxicity toward military ends, new regulation of the workplace sought to hold toxicity within certain prescribed levels of acceptable exposure. Here, the environment was brought into political being not as a weapon but as an administrative domain that might better contain toxicity.

Like so many other stories of our present, much of this began in the factory. It was here, as historians like Christopher Sellers (1994, 1997) and Michelle Murphy (2006) have documented, that a new form of scientific expertise took shape around toxic exposures in industrial production that previews many of our contemporary environmental protections. "Modern war," an oft-repeated newspaper slogan summarized during World War I, "is largely a matter of chemistry and engineering" (Herty 1916: 4821). Such chemistry was at once vital to the war effort and fraught with medical misgivings in its manufacture.[32] As workers were sickened in munitions factories, refineries, and petrochemical plants, rising incidents of "industrial disease" moved to the forefront of the federal agenda as the US Department of Labor authorized new interventions into factories as a part of urgent efforts toward "conserving industrial manpower" during war. Alice Hamilton (1919: 248), one of the first doctors commissioned by DOL to study the "dangerous trades" during World War I, placed six medical students "well trained in laboratory methods and in making clinical observations" of workers inside munitions factories.[33] Embedded on shop floors for "one to two months," there was a concerted effort to understand the ailments that afflicted workers in a more methodical and precise manner (Hamilton 1919: 248).[34] These investigations helped give rise to industrial hygiene, a new medical science that placed toxic exposures in the workplace at the root of a peculiar family of ailments. During World War I and in the years after, industrial hygienists came to usurp the role of the factory physician and the union health clinic and inserted a new form of medical expertise whose authority rested on its independence from labor and capital.[35]

As industrial hygiene developed, its focus turned from the clinical inspection of workers' bodies to the technical monitoring mediums of exposure. Armed with new chemical detection devices, industrial hygiene transformed factories into an experimental field within which specific "industrial poisons" could be objectified in the air and water for more exacting analysis and administration.[36] For industrial hygienists, the safety of the workplace was achieved not through staking out a political position on toxicity or through advocating for a certain class of people but by determining the line at which key industrial ingredients became dangerous industrial diseases. This was accomplished largely through defining an empirical boundary between safe and unsafe concentrations of specific chemicals and then assembling devices to monitor those thresholds inside factories, whether as "toxic limits" (Schereschewsky 1915),"safe concentration" (Sayers, Meriwether, and Yant 1922), "maximum allowable concentrations"(Cook 1945; Elkins 1948), or "threshold limit values"(Coleman 1955). Such thresholds, as historian Christopher Sellers (1997: 2) demonstrates, offered a crucial precursor of environmental governance, as they were "the first to tabulate lists of threshold concentration levels, and the first to devise the kinds of precise delineations between the normal and abnormal that underlie today's environmental law and policy, as well as its science." These boundaries, unaligned with class interests and presented as biochemical fact, came to inform factory design as insurance companies, trade associations, and municipal building codes took them up as enforceable guidelines. This analytical stabilization of the factories' interior provided a novel means of holding industrial chemicals and workers' health at arm's length and policing the mediums, like air, that brought them into consequential contact. Displacing a long-standing point of confrontation between labor and capital, the science of industrial hygiene helped transform the politics of working conditions into a simple matter of compliance.

Industrial hygienists sometimes described how they would find potentially dangerous factories in unfamiliar cities. They would look up and follow the telltale smoky emissions to the source. The engineered fix to workplace toxicity, as historians like Joel Tarr (1996) have shown, was simply to vent the problem out of the factory. As environmental engineers often say, dilution was the solution. This, of course, did not so much solve the problem as displace it. Having mastered chemical afflictions inside the factory, industrial hygienists soon found themselves in the homes of workers, where children and neighbors suffered similarly without having ever stepped into the factory. As industrial emissions drifted into adjacent neighborhoods, industrial hygienists were called to the scene, first for the fog disasters that seemed to plague industrial cities in the 1930s and 1940s and then within municipal governance in the 1950s and 1960s, working to rein in what became known as the pollution problem. Los Angeles, Detroit, Denver, Pittsburgh, New York, and other large cities hired teams of industrial hygienists to help hold air pollution within certain limits. Most prominently, industrial hygienists helped establish urban thresholds for carbon dioxide, sulfur oxides, soot, ammonia, nitrogen oxides, aerosols, ozone, hydrocarbons—and sometimes radiation and pollen. Nearly all of these chemicals, it should be noted, have a single source: they are by-products of fossil fuel combustion.

The devices and disciplinary practice of industrial hygiene helped make petro-pollution visible within the city limits. As in the factory before, managing pollution was premised on first creating a field of scientific legibility against which the problem of pollution could be seen in a more objective light. Sidestepping the specific sources of toxicity like actual smokestacks, petrochemical plants, and automobiles, governance turned instead to stabilizing the mediums of exposure. Here, the management of urban pollution came to hold the public and industry at arm's length and began policing the air and water that might bring them into consequential contact. And again, as in the factory before, this was effective to the extent that a direct confrontation between citizens and industry was rendered difficult if not impossible.

The history of toxic exposure is also a history of analytic containers. In some ways, the emergence of "the environment" is the story of how the biological reach of petro-pollution came into focus outside the built mechanisms of control like the factory or the city. In the 1960s and 1970s industries eluded regulations by designing bigger smokestacks, flushing waste downriver of the city, or simply building new plants just beyond municipal jurisdiction. As lakes were declared dead, rivers caught fire, and mountains were shorn of vegetation as rain turned acidic, the effects of diffuse pollution became a rising national crisis. At the same time, there was a growing realization that the impact of fossil fuels emissions was not confined to the place of combustion and that the impact of petrochemicals was not limited to the place of application. As both started showing up in lung tissue, blood samples, bird eggs, farm produce, lakes and streams, and atmospheric systems, there was recognition of the sprawling negative ecologies of hydrocarbons. Here, thresholds offered a novel means of sidestepping any reckoning with these ecological webs and their radical implications. Thresholds turned attention exclusively to the mediums of exposure like air and water, treating them as autonomous fields in which pollution could be measured and managed without bothering with the material soure. As researchers working for Ralph Nader noted in 1970, "The largest single source of air pollution, the automobile, was never mentioned in federal legislation" until the late 1960s (Esposito 1970: 22).

Following the model of the factory and the industrial city, the turn to the environment established a national jurisdiction for the implementation and enforcement of thresholds. By the 1960s, pollution was polling second only to crime as the greatest threat to American well-being (Markowitz and Rosner 2002: 155). In 1969, less than 1% of Americans prioritized the environment. Two years later, a quarter of all Americans believed protecting the environment was of crucial importance (Whitaker 1988). At first this formed a crisis without a constituency, a lack both Democrats and Republicans were eager to amend.[37] In a flurry of one-upmanship, Democratic leaders in the Senate shepherded two major expansions of federal power into law—the Clean Air Act (1970) and the Clean Water Act (1972)—while President Nixon consolidated the tasks of enforcing these nationalized definitions of air and water quality into an emboldened and strikingly unbeholden new agency: the Environmental Protection Agency (EPA).[38] Both parties clamored to claim the environment.

In 1971, the federal government established enforceable national air standards (with a new enforcer, the EPA) for five pollutants, all of them emissions from fossil fuels: sulfur dioxide, particulates, hydrocarbons, carbon monoxide, and photochemical oxidants. Initially, these standards attempted to balance historical averages of emissions with health concerns, but a number of lawsuits compelled the EPA to privilege health concerns (Jones 1975). As investigations began to show the chronic harm from even low-level exposure, federal standards for each of these pollutants were quickly ratcheted down. By 1973, there was growing debate about whether hydrocarbon emissions should be tolerated at all. Under pressure from citizen lawsuits and the courts to let science dictate policy, the EPA called for drastic reductions in fossil fuels use in seventeen states and major cities like Denver, New York, and Pittsburgh. The EPA suggested these cities build mass transit systems and start rationing gasoline to bring their air quality into compliance with the national standard. Los Angeles was ordered to reduce gasoline use by 82 percent during summer months (Jones 1975: 270). In Philadelphia and Pittsburgh, city leaders were told to remove 200,000 cars from the road. By the end of 1973 and with the OPEC embargo looming, President Nixon stepped in. Addressing the nation, Nixon (1973) demanded that Congress provide him with the exceptional authority "to relax environmental regulations on a temporary case-by-case basis, thus permitting an appropriate balancing of our environmental interests ... with our energy requirements, which, of course, are indispensable." This confrontation quickly formalized around crude oil and vertically ranked what were now two entirely separate technical properties of petroleum: an external science of gain (the economy) and an internal science of harm (the environment). Whether by the coercion of the US-led developmental loans or by the elective choice of national leaders, such thresholds soon became the basis of environmental governance in nations across the world. As in the United States, the vast majority targeted the emissions of fossil fuels.

ASSESSING THE IMPACT

In 1963, a professor of government at Indiana University penned an essay titled "Environment: A New Focus for Public Policy?" (Caldwell 1963). The paper, circulating widely in legislative circles in Washington, DC, was disarmingly straightforward: in response to a growing crisis of life, the basic conditions of life should be administered as a distinct federal domain with its own institutional apparatus. Rachel Carson's *Silent Spring* had been published the year before, causing a firestorm of concern about the disconcerting and disruptive reach of new petro-chemicals into the fabric of life. Amid a tidal wave of interest, this essay argued that, properly conceptualized, the environment should offer tri-age for the worsening conditions of life.

Lynton Caldwell, the author, was soon invited to DC to draft the first federal environmental policy, the National Environmental Policy Act (NEPA) of 1969. Caldwell's reflections on crafting this policy and on the wider necessity for environmental governance offer a window into the conceptual and administrative shift the environment entailed.[39] In response to growing evidence that the "life support capabilities" of the planet were at risk, Caldwell (1998: 5) later reflected on how politi-cal scientists, lawyers, and biologists gathered in Washington, DC, to "reconceptualize the environment in relation to the responsibilities and functions of governance." The reverse was true as well, as many elected officials worked to upgrade the capacity of the federal government in response to the growing recognition of the environment.

As "the introduction of chlorinated hydrocarbons and radioactive isotopes into food chains" so effectively exemplified, Caldwell wrote (1970: 82), the present was beset by threats that did not abide by exist-ing jurisdiction or bend to inherited wisdom. For Caldwell, this crisis of life demanded a new infrastructure of governance. The former rubric of managing the natural world—the conservation of natural resources—was ill-equipped for the present crisis. The point was not, Caldwell wrote, to preserve some "ecological islands" among wider "biophysical ruin." Conservation was too human centric, too reliant on industrial reasoning, and too, well, conservative. What was desperately needed was not just to safeguard future extraction or protect isolated areas but to stabilize the conditions of life itself. Early attempts at this stabilization were proposed at the level of rights. An early draft of NEPA stated: "Each person has a fundamental and inalienable right to a healthy environment." As early as 1966, Caldwell himself had broached the idea of instilling "public (or private) rights in environments-as-such" (659), that is, making the environment a rights-bearing subject itself. (This was not as far-fetched as it may sound; Supreme Court justice William O. Douglas's 1972 dissent in *Sierra Club v. Morton* pointedly raised the possibility of granting the environment the same rights as a corporation.)[40]

Unable to overcome questions of how such rights would actually work, and somewhat smitten with the rising role of economics in influencing federal policy, another position won out. Instead of an expansion of environmental rights, NEPA worked to interject environmental expertise into the everyday functions of governance. With the viability of earthly life hanging in the balance, the management of the environment, as Caldwell (650) had written in 1966, had to move beyond democratic debates to make those decisions "that a society knowledgeable of its own needs, interests, and potentialities *ought* to make." This, for Caldwell and the authors of NEPA, involved explicating the contingencies of life and bringing that knowledge into high-level decision-making. Cultivating such enlightened decision-making was actively contrasted to that other genealogy of the environment: thresholds. While thresholds provided an "external policing mechanism" to protect the environment, NEPA aimed to "internalize" the environment within federal decision-making (Liroff 1976: 18–19). In a division of labor Michel Foucault would have found fitting, thresholds threatened punishment, while NEPA disciplined from within.

Modeled on how the "the economy" had come to inform, orient, and discipline policy independent of democratic deliberation, NEPA sought to introduce scientific knowledge of life's precarious balance into every aspect of governance (and, as with the economy, sought to do so not on the shoulders of popular sovereignty but by equipping technical expertise to override democratic practice). Parts of NEPA were literally copied and pasted from the Full Employment Act of 1946, which introduced "the economy" into federal governance and brought a new cadre of economists into the White House in the form of the Council of Economic Advisors.

As Timothy Mitchell (1998), Donald MacKenzie (2008), and Koray Cal-
iskan and Michel Callon (2009) have all argued about the economy, it is
often the methods of fact production and genres of interpretation that
instantiated the socio-material field (not the other way around). Here too,
the methods and genres of the environmental impact assessment came to
instantiate the socio-materiality of the environment itself.

If the economy introduced a scientific regime of scarcity into governance, the environment introduced a scientific regime of vulnerability.[41]
Even as the economy and the environment have come to stand in stark
opposition to one another, they should be understood as mirror formations: each produced an expertise whose authority was realized in its
achieved distance from embedded and embodied knowledge, each formatted life for state rule, and each sought to discipline the present according
to its modeled vision of the future. (This conceptualization leans heavily on Timothy Mitchell's (1998) incisive archaeology of "the economy,"
even as it should be noted that "the environment" was never able to shake
the weight of materiality in the same way the economy was.)

NEPA advanced two methods of bringing this new science of vulnerability into governance: a new presidential council of environmental
experts and, in what is likely "the most imitated U.S. law in history," the
environmental impact assessment (Yost 1992: 6). The former brought
new environmental expertise into the White House, while the latter distributed a new kind of environmental calculus to each and every govern-
mental project.[42] The environmental impact assessment was described
by Caldwell in 1966 as the "drawing up of a balance sheet of ecological accounts by which the true costs and benefits of alternative decisions might be compared" (524). This novel environmental ledger has
since become a ubiquitous technique and cultural icon, mocked in *New
Yorker* cartoons and late-night quips about the nanny state. Yet its
intervention should not be discounted. It opened previously invisible
decision-making processes and their implications to public inspection
(while limiting what the public might do about them), and soon became
a bureaucratic lever upon which enormous fortunes might rise or fall.
It also conjured a new consequential field of fact production, modeled
futures, and institutional morality.

As envisioned by Caldwell, the environmental impact assessments
would usher in a new regime of "surveillance" that would produce
untold amounts of data on the conditions of life.[43] Taking stock of the
likely effects of a project on nearby water quality, air quality, species habitats, ecosystems, and more, such impact metrics required an objective

definition of normal as the baseline against which potential disruptions could be measured and managed. Impact assessments would, in Caldwell's words, "establish ecological baselines—parameters, ranges, and gradients for sustaining life" (1970: 84). Not only would these baselines help determine if a project should go forward, but the resulting knowledge could also help manage those projects as they unfolded, providing an early warning detection system if things started to go awry, as well as a road map for restoration if needed. Instead of coming up with a universal or national baseline of life, NEPA distributed that task to the specific locales where disruption was anticipated. Each new project would require its own accounting of the localized conditions of life it might infringe upon, its own project-specific definition of vulnerable life. Crucially, such work does not unify an official understanding of normal life, but provokes a proliferation of normative definitions. Each new project reifies the constitution of normal life in its shadow.

In 1970, this "major revision of the administrative functions of the U.S. Government," as the lone dissenting voice in congressional debates bemoaned, passed nearly unanimously, without much fanfare, and with little appreciation of the minor revolution in bureaucratic procedure it was instigating (Liroff 1976: 30). Few perceived the sheer breadth of its implications or anticipated the entire industries of environmental law and consulting that would take shape in and around its guidelines. Indeed, interviewed years later, many key participants shared the sentiment of one congressional staffer who commented, "If Congress had appreciated what the law would do, it would not have passed" (Liroff 1976: 35). In the years since, the environmental impact assessment has proliferated worldwide as a basic tool of governance in cities, industries, nations, and international organizations. As Michael Watts (2005) and others have observed, perhaps no industry has become as heavily invested in environmental impact assessment as the fossil fuel industry.

ENVIRONMENTAL SCIENCE AND POLICY

Over the past fifty years, the environment has taken forceful shape around two instrumental genealogies: thresholds of toxicity and environmental impact assessments. Both emerged in response to the negative excess of American power, and both initially promised radical new forums to reckon with and rein in that excess, whether by privileging health concerns or by emboldening rights. Yet as they developed into federal policy, both turned their backs on the negative ecologies of American power as they worked to stabilize the effects as an entirely separate matter of concern. Thresholds and impact assessments proved key to this tempered detachment. As they became taken up by the state, both worked to stabilized a new field of objectivity and operation within which the new disciplines of environmental science and environmental law took root. The negative ecologies of fossil fuels and nuclear weapons—as a unified web of synthetic destruction—drifted out of focus as the environment rendered the fallout a fractured family of autonomous fields of fact production and regulatory compliance. A wounded world was given new coherence not by the materiality of its injury but by the methods for its accounting. As mediums of exposure and fields of impact became premier field laboratories for the specialized measurement and management of harm, the underlying materiality of fossil fuels was evacuated except for its outlying effects. Such work can be ruthlessly proficient, and instantiations of the environment along these lines have been instrumental in saving lives and reducing pollution worldwide. Yet by turning their backs on the negative ecologies of fossil fuels, the possibility of a more fulsome confrontation with the underlying cause of the crisis of life became exceedingly difficult to imagine, let alone enact.

These thresholds and impact assessments also underwrite the normative criteria for environmental critique. Environmental protections displace a politics of confrontation as they push effective action against fossil fuels into the realm of standardized methods, certified results, acceptable levels, and codified assessment models. This has serious consequence, for not only does the environment divorce measures of harm from measures of gain, but the category has found its most forceful definition through moralizing and managing an ahistorical, moderately contaminated, and exceedingly technical understanding of life.

Yet the environment was never fully contained by a single expertise. Unlike the economy, no single discipline or institutional authority gained priestly privilege over the environment. Even as the environment took forceful shape around thresholds and impact assessments, these instrumental technologies were never fully monopolized. Their technical application formed its own political field at the intersection of agency hubris, corporate interests, and organized advocacy. While the environment may have displaced a politics of confrontation, it did not render the environment a purely apolitical forum: conflict and rigorous debates continue to shape the content and implementation of environmental protections. Environmental science, as so many of its practitioners readily admit, is also a political practice.

Although it has long shed the revolutionary possibilities that attended its conception, environmental science has not shed its public relevance. Environmental science took root in the state recognition of the environment, and has become one of the most substantial engines of contemporary fact production. The regulatory enshrinement of thresholds and impact assessments opened up new analytical fields of research and publication, new disciplinary associations of expertise and employment, and new applications of scientific facts for matters of law and liability. The birth of environmental science, writes historian Samuel Hays (2000: 137), "is one of the most significant developments in science in the last half of the twentieth century." Even as environmental science has become massively instrumental to state policy and extractive projects, it so often remains a derided topic of scholarly engagement. Immense fortunes now ride on the findings of environmental science, as does the drafting and implementation of huge swathes of local, national, and international law. Fielding contributions from academics, state agencies, corporations, and consultants, environmental science journals can have far greater readership than flagship disciplinary journals and impact factors that go far beyond scholarly citation. Yet to our great misfortune, we primarily understand environmental science through how it lacks the prestige of pure academic science. Environmental science has real purchase in the world, yet that overt worldliness is often the very reason many dismiss it. The disinterested pedestal of academic disciplinary science remains paradigmatic in our understanding of the practice of science, to the great impoverishment of our grasp of the political field of environmental science. Historians have drawn attention to the way universities became a new locus of scientific production in postwar America, pulling the laboratory away from the nuclear state and raising the status of fact production in relation to its disinterest (Egan 2007: 25). Such an account, however, sometimes misses the shift of fact production back to government, industry, and above all consultants post-1970 (scientific education still unfolds almost within the university, but the production of certified knowledge by science is increasingly outside the university system). The rise of environmental science exemplifies this shift. Many environmental scientists find lucrative work in the oil industry monitoring compliance with thresholds and preparing impact statements, while the EPA has become the "largest civilian arm of the US government," and state-level environmental agencies are some of the largest public employers in their states(Guha 2000: 83). Environmental science unfolds not within the relative institutional autonomy (and political insulation) of the academy, but within government agencies, corporate offices, courtroom litigation, and a veritable army of consultants. Environmental science "begins with the assumption that human health and industrial poisons can co-occur," write Kroll-Smith and Lancaster (2002: 204); it is premised on the notion that the environment and bodies "are sufficiently disparate" as to allow a quantitative science to unfold in the space between them. However questionable its founding assumption, environmental science nonetheless retains a fierce sense of analytical ethics, an objectivity forged not from denying politics but from acknowledging it. Environmental science has never been disinterested; the robustness and effectiveness of its produced knowledge comes far more from being pinned down in a cross fire of interest than from somehow imagining itself to rise above the swirl of baser motives. Environmental science has developed into a science comfortable within the mess of the world. It does not aspire to purity, whether in the disciplinary insulation of inquiry or as an empirical goal of research, but roots itself in the necessity of a factual basis of action. It is, as Samuel Hays (2000: 151) has described, "a decision science." Yet there is a political struggle at the core of the field, one centering on the direction of application: Is the environment a field of science that provides the secular ethics of extraction or a science that insists on the priority of care in a broken world?

The political conflicts at the core of environmental science unfold almost entirely upon the terrain of technical legibility authorized by thresholds and impact assessments, and rarely reach for a more direct confrontation with the historical materialism of the cause. As these conflicts have proceeded, they have deepened the technical qualification of environmental protections, abstracting them further and further from the manner in which people actually live the problems they regulate.

The environment also exceeded technical expertise in another way. The underlying problem, both in the open-ended ecological harm unloosed by the imperial rise of America and in the movements that gathered around the eroding conditions of life, routinely exceeded the fixtures of technical control. Even as thresholds and impact assessments engineered the authoritative objectivity of the environment, the fallout of American power continued to defy such enacted stability and partial measures. The negative ecologies of fossil fuels were never fully captured by the format of the environment, as so many living downstream of extractive projects, on the fence lines of petrochemical plants, or near where the infrastructure breaks know all too well. Their voiced protest in the lived dimensions of pollution often demonstrate the impoverishment of thresholds and impact assessments. Yet rather than countering the technical foundation of the environment, the felt parameters of pollution often end up working to give new ethical momentum to administrative techniques aimed at objectifying vulnerable life. Gaining any traction in environmental protection requires first disciplining experiences of profound harm into the acceptable measures of harm. The felt shortcomings of the environment paradoxically bolster its legitimacy, widen its reach. This disjuncture has become the experiential field of injustice around fossil fuels, dictating the manner in which people live under the weight of both negative ecologies and the official legibility of their effects. The ethnographic chapters that follow inhabit this disjuncture—its barricaded possibility, the complicity it requires to begin work on incremental improvement, and the rage that still overflows—across the contentious sites of fossil fuels in contemporary North America. It is no coincidence that the rise of environmental science mirrors the consumption of fossil fuels in the United States. And it is also no coincidence that environmental science has not so much checked the addiction to fossil fuels as provided acceptable parameters for it to deepen and expand.

The crisis we face today is not that the United States monopolized these techniques of synthetic might and institutionalized blinders, but that America pioneered them.[44] "Oil civilization began in the USA," John Urry (2013: 10) writes, and if the resulting American Dream becomes the global measure of the good life "it would take at least five planets to support it."

DECOLONIZING THE ENVIRONMENT

It could have been different. As mentioned previously, prompted by an announcement from the secretary general in 1969 that "it is becoming apparent that if current trends continue, the future of life on Earth could be endangered," the UN organized a conference on the human environment in 1972 (cited in Kennan 1970: 191; Ward and Dubos 1972; UN 1972). Initially, the conference had the aim of authorizing planetary thresholds and routinizing environmental impact assessment at a global level. A very different debate unfolded in Stockholm. Representatives from the rest of the world agreed that there should be a universal human right to "adequate conditions of life" but insisted that threats to life have both biochemical and historical roots. Such a move brought American empire into disconcerting focus. "In this respect," notes the conference report, "policies promoting or perpetuating apartheid, racial segregation, discrimination, colonial and other forms of oppression and foreign domination stand condemned and must be eliminated" (Ward and Dubos 1972: 4) as part of environmental protections. Here, "a world-wide harmonization of standards" (26) and a planetary network of "baseline stations" sits uneasily next to a political accounting of the ecological debts of racism, colonialism, and underdevelopment. Under the umbrella of "the environment," many representatives at the UN conference insisted on "a new liberation movement" to confront toxic pollution *and* the ecological effects of empire as a conjoined threat to the conditions of life on Earth. While the UN added a Program on Environment with headquarters in Kenya soon afterward, its work drifted away from the promise of a historical reckoning and soon came to focus on widening the reach of thresholds and impact assessments.

POSTSCRIPT: HISTORY AS ANTHROPOLOGICAL THEORY

This chapter has drawn historical attention not only to *what* we know of the environment but also *how* we have come to know the environment. Just as the term was being enlisted to govern the crisis of life in the 1960s, Eric Wolf (1965: x) described the environment as a "watertight" concept compared with the stress fractures emerging in the concept of culture as that core concept of American anthropology found itself pulled into consequential public relevance. Wolf, perhaps, was a bit too retrospective with the environment, looking toward the deep intellectual roots of the term, not its rising political scope. As an erudite emphasis on the influence of context, environment—US social science's first translation of Auguste Comte's *milieu* and Wilhelm Dilthey's *Umwelt*—has a rather distinguished intellectual pedigree.[45] Perhaps more akin to culture than Wolf realized, in the 1970s the scholarly problem soon became one of a concept coming to matter a bit too much. The ecological crucible described in this chapter, and the instrumental genealogies brought to bear on it, drastically recast this rather highbrow academic frame as the technical field that could best regulate the negative excess of modern power.[46] As the ethical and political action oriented by the environment exceeded its scholarly purchase, the technical authority of the concept rose in inverse relation to its intellectual foundations. It's coherence came to lie far more in the practical consequence of its application than in its empirical content or theoretical definition.

As it was taken up worldwide, the environment also came to displace efforts to center the unbound crisis of petro-capitalism and thermonuclear statecraft as it shifted attention instead toward equipping states to manage their effects. This shift, burying the interwoven materiality of the crisis underneath the disciplinary fields of calculating and administering effects, mirrors a trend in social theory. For this moment of history also heralds the beginning of a broad turn in social theory toward the preferential treatment of the symbolic, the interpretive, and above all, the discursive, in the ordering of social life. Materiality, within the terrain of state administration and social theory, was pushed into an ancillary effect. That project, one that formed a good part of my graduate education, would focus attention on the realities incarcerated by various forms and forums of objectivity. The effects of objectivity marked the starting point of theoretical significance and ethnographic description in the anthropology of science. While hugely generative, such a stance too easily dismissed materiality as secondary and science as ideology. As historians John McNeill and Peter Engelke (2016: 210) have written of the cultural turn in the social sciences and its studied avoidance of ecological crisis: "The intellectual flight from reality made it slightly easier for those in positions of power to avoid facing up to it." Writing in 2002, Mike Davis put a slightly sharper spin on it: "Although the academy may still favor the esoteric relativity of postmodern textualism, vulgar economic determinism—which begins and ends with the superprofits of the energy sector—holds the real seats of power. We don't need Derrida to know which way the wind blows or why the pack ice is disappearing" (417).

The displacement of materiality shaped many ethnographies of communities living with the toxicity of fossil fuels. Many monographs either worked to make frontline communities' suffering more legible to the modular objectivity of thresholds and impact assessments, or they worked to downplay those instruments of the state in order to privilege the subjective experience of contamination over any field of material reckoning. In both cases, the biochemical routes and accruals of harm drifted into the background as the objectivity of thresholds and impact assessments served to frame the significance of environmental ethnography and its pursuit of justice, whether by tactical inclusion or theoretical omission. The environment, as a historically constructed scientific and legal field of fairly recent vintage, remained largely unfazed: it remains the main stage upon which various local, moral, and disciplinary dramas played themselves out. As an active force, materiality remained backstage.

A striking thing has happened in the past few years. Materiality has come roaring back to life, and with it a new appreciation for the life force of the nonhuman and the wonder of science. This current of materialist thought often reduces the significance of the past to examples of a flawed and failed epistemology. Among other things, this flattens all questions of historical contingency and political struggle into accounts of uniformly bad philosophy. The truly critical task, for this current of anthropological thought, is to properly grasp the liveliness of the material world unadorned with modernist form, an emergent apprehension that promises to create the world anew. Curiously, this resurgent materiality in theory and research now accuses the environment with preventing just such a revolutionary understanding. For many scholars of this moment, the environment is now coming into focus as the impediment that long blocked more emancipatory insurgencies. Environmental protections, write Jason Moore and Raj Patel (2017: 40), "rest on the historically bankrupt idea of immutable human separation from nature." This nature/society dualism, they argue with many others, stands both complicit and condemned in the face of planetary crisis. The environment, writes Kregg Hetherington (2019: 5), inferred that "problems and the people who suffered from them could always be localized." The environment implied there was "an intellectual outside" to the problem, that could then inspect problems as a contained phenomenon. "The conceit of Anthropocene," Hetherington writes, has no such outside. "In place of 'environment,' there is now the Earth system," argue Bonneuill and Fressoz (2016: 20), as the Anthropocene fundamentally breaks with the technical modalities of the environment.[47] Again and again, the impoverished constitution of the "environment" provides a point of scholarly departure to finally give our contemporary upheavals a more epochal definition.

In many respects, these scholars are undoubtedly correct. The environment severely constrains our view of our present crisis. Yet the resulting sense of scholarly revelation is premised on bad history. The negative ecologies of petro-capitalism and thermonuclear statecraft that came into focus in the 1960s and 1970s defied any simple division of nature and society. Rachel Carson, Barry Commoner, and others strove for a new analytical and political vocabulary beyond such impoverished dualisms, dualisms they too found dangerously complicit. Even as the environment displaced the radical implications of their negative ecology, it did not return to these dualisms. Thresholds and impact assessments do not reify the boundaries of nature and society so much as manage their overlay.

Nor were the founding problems of the environment considered individualized or insular. DDT and strontium 90 were not "localized" problems by any stretch of the imagination. As Commoner wrote in 1967: "The new hazards are neither local nor brief" (28). It was clear to scientists, citizens, and state officials at the time that the afterlives of fossil fuels and nuclear weapons defied every existing jurisdiction. Nor does the Anthropocene represent a straightforward break with the environment. Many of the approaches being advanced to hold back the worst of climate change, from "planetary boundaries"(Rockström et al. 2009) to the Intergovernmental Panel on Climate Change's (IPCC's) emissions scenarios (2014, 2018) seem not to break with toxic thresholds and impact assessments so much as to deepen and widen their scope beyond the state.

The environment may very well be the conceptual barricade that prevents our owning up to the present crisis of life and prosecuting the profiteers of destruction. But it should be recalled that the environment was also the engineered solution to just such a revolution. The resurgent materialist theories in social research often presume that if only we might bring our thinking in alignment with the agencies beyond the human, a new politics would become possible to free us from this bind. The formation of the environment in the 1960s and 1970s seems an instructive case to temper such promise, both in its founding attention to negative ecologies of modern power and in how radical possibilities of that science were so quickly enlisted into an overwhelmingly technocratic solution. It's curious to see how the crisis of life that provoked the environment—a history I have sketched out in this chapter— carried so much of the vital materialism, planetary orientation, and clamor over a new ethical and political vocabulary that is now being taken up today as proof of our unprecedented condition. At that earlier moment, an emerging crisis of life brought into sharp focus an understanding of petro-capitalism coursing through the veins of an ecologically interconnected planet as life veered toward rampant destruction. Those earlier claims were read widely and provoked new policy and public awareness. Even as they gestured toward revolutionary resolution, such popular insights were soon given an institutional definition that domesticated the negative excess of power as a side project of the instrumental reason of the state. My sketch of this history is less a finished project than a preliminary effort to brush away the accredited nonsense clamoring to contain the frontline laments of contamination. My provisional turn to this history of our present aims to give empirical and theoretical credence to their complaints. Such work is not aimed at getting away from the present but at providing new coordinates for ethnography to come closer still to the ecological crisis of now.

Instead of privileging the possibilities of materialist theory over historical realities, we might return to the flawed and fatal ground upon which we still live. Besieged by war, toxins, climate, and pandemics, negative ecologies of all variety assail our present. Attention to the materiality of these widening webs of injury tempers the ontological optimism of contemporary anthropological theory and insists on explanations adequate to those lives bruised by the worsening condition of the world. Theory should bring us closer to the world at hand, not with an aim to naturalize its forms but to better understand the contingencies of those forms and the manner in which they might be overcome now. Possibility is not the unique property of utopian futures. The historical grounds we inhabit—its haunting insights, blind alleys, and lost causes—also advances anthropological theory for a better world today.

CHAPTER 2

Governing Disaster

One year after the BP oil spill, I followed a caravan of federal officials as they toured the Gulf Coast and explained to local communities how damage to the environment would be assessed and ultimately righted. During these meetings, federal officials fulfilled their mandate to elicit public input on how to fix the environment after a major oil spill. Held in high-school gymnasiums and community centers, these meetings were peculiar affairs at which industry lawyers, environmental nongovernmental organizations (NGOs), and municipal leaders melded with angry residents of a "public" lined up in rows of folding chairs.

Each presentation began with the same line: "First and foremost this was a human tragedy. Eleven workers lost their lives," before pivoting to the point: "But tonight we're going to focus on the environment." The goal of these meetings and the interventions they foreshadowed was explicit: "to make the environment whole again." These meetings routinely concluded with federal officials exhorting audience members to express their vision of an environment restored to normal: "We don't have a manual for how to put everything back together again. We need your help, your input. Are there specific species you want to make sure we pay attention to? Are there specific sites you would like us to focus on?"

An interested public hewed close to this script. At one meeting, a number of suited men read from a memo entitled "Talking Points for Environmental Restoration," authored by an industry group. Each reiterated the same project: scuttling old oil rigs to form artificial reefs.

Representatives of environmental NGOs urged action on various topics close to their organizations' missions: endangered species, wilderness areas, hunting and fishing, or ocean conservation. Equally present were municipal and state officials, pitching projects like sprucing up a waterfront or building a boat ramp as key components of environmental restoration. Most of the public input struck me as pointedly private, consisting of various long-standing agendas repackaged to be newly persuasive in the coming bonanza of restoration.

There was also a different, more dissonant public at each meeting: sickened residents. Their voices, unruly and unvetted, offered a far messier and almost nightmarish accounting of the oil spill. One woman interrupted a meeting by handing out lab reports on her blood: "I have poly-aromatic hydrocarbons in my blood. I need help." A middle-aged women rose in the back and moved to the microphone: "The water is poisoned. We're weren't involved in the oil industry or the cleanup effort, we just live here. I have trouble breathing now. We need medical care." One doctor stood and introduced two of his patients: "These men are extraordinarily ill. The oil was in the water and now it's in our blood," he said. "Feel free to question them." For these residents, the imperiled ocean stretched into their bodies. Together, their voices offered an unsettling refrain: we live and work and eat in ways that confuse any hard and fast distinction between an environment and a public. Frustrated by official evasion, one woman asked: "Do you not think the health of environment is related to the health of the residents?" Another woman said: "The water and air are poisoned. The environment is killing us." Federal officials dismissed these comments with the same polite recusal: "This is a meeting about damages to the environment. Your concerns are best addressed elsewhere."[1] Occasionally, officials would discreetly pass these residents a brochure entitled "Mental Health and the Oil Spill." I asked one official charged with formatting public input at these meetings what she did with such comments. "Nothing," she said. "They don't fit."

Such fraught scenes articulate the emerging categorical fault lines between the public and the environment that underlay the official response to the BP oil spill. In this chapter, I describe the epistemic politics of the environment during this oil spill, its contested boundaries, and forceful enactment. This chapter does not begin after the fact of the environment but focuses squarely on the making (and remaking) of that consequential domain during a major oil spill. Disasters do not unfold within the disciplined fields of knowledge or the settled jurisdictions of

FIGURE 3. Gulf Coast during the BP oil spill, Grande Isle, Louisiana, 2010. Photo by author.

governance. "Oil spills," I heard Admiral Thad Allen quip at the height of the BP oil spill, "are agnostic to political boundaries." Disasters do not abide by the working partitions of research and rule; they establish those distinctions anew (Fortun 2001; Petryna 2002). During the BP oil spill, the environment came apart and then was put back together as the experimental domain within which this unprecedented oil spill could be objectified for both scientific quantification and political responsibility (Porter 1995). Following Rosalind Shaw (1992: 200) on the anthropology of catastrophe, "the question here is not what *causes* events but their *conceptualization*."

FIXING THE ENVIRONMENT

The BP oil spill defined a vital frontier of knowledge. Unfolding nearly a mile underwater and beyond the pale of easy observation and easy capture, it overwhelmed established understandings of both oil spills and the ocean's vulnerability to them. As we now know, less than 10 percent of this shockingly large deepwater blowout rose to form a surface slick (NOAA 2010). The vast remainder of this oil spill—roughly fifteen *Exxon Valdez*es unfolded within the ocean itself. The immense pressure of the deepwater (and the added force of chemical dispersants) broke down the crude oil into its component parts. These parts, in turn, had distinct trajectories. Gaseous hydrocarbons like methane, by far the largest component of this oil spill, dissolved into the ocean or formed tiny ice particles infinitely suspended at varying depths within the water column (Kessler et al. 2010; Valentine et al. 2010). Swept into subsea currents, they formed underwater plumes of methane-laden seawater (Camilli et al. 2010; Joye, Macdonald et al. 2011). Heavier hydrocarbons sank to the ocean floor, coating it in a thick tar (Joye, Leifer et al. 2011). Many of the aromatic hydrocarbons, like benzene, rose to the surface and quickly evaporated into the atmosphere, creating ephemeral trails of carcinogenic air that drifted over parts of Louisiana and Florida (de Gouw et al. 2011). The hydrocarbons of this spill were legion. Not only did this render the coordinates of the oil spill multiple and frighteningly unbound, it also meant its biological consequences were occurring on the outskirts of the forms of life the state had historically sought to protect under environmental law.

During the BP oil spill, "the environment" in need of protection was very much an open and urgent question.[2] Scientific and political consensus on what counts as "the environment" during this oil spill was not the starting point of the official response but its consequential outcome. Taking the environment as a compelling ethnographic question, I describe the situated debates and novel technologies that sought to bring the microbial and deepwater dimensions of this oil spill into a working correspondence with the historical dimensions of the defendable environment. This process is significant not only for how official knowledge of an oil spill is produced and validated but also for what is left out. The limits of pollution (and its effects) have come to rest not on the outer edges of felt impact, human or otherwise, but on the legibility of such claims within the present constitution of the environment.

The environment, one White House adviser told me in a rushed conversation between meetings at the height of the spill, "is not something you can put a fence around and save." No, she said, the environment is about a proper understanding of "ecological functions" and what threatens them. The urgent "knowledge practices" that coalesce around the governance of the environment, to which both Kim Fortun (2001: 7) and Timothy Choy (2011: 5) have drawn our attention, are oriented not toward matters of scarcity but toward problems of vulnerability. Here, the focus is not so much on how environmental issues play out within a larger political economy as it is on the analytical operations that stabilize endangered life as a legible object for uniform measurement and centralized administration.[3] The environment that was twisted and pulled into novel application during the BP oil spill was not a strategically distributed natural world, but rather an urgent scientific project to objectify the immediate vulnerability of life for effective governance.

Environmental protections, as I argue here, are not the safeguarding of an obvious place but rather an "epistemic habit" that forcefully instantiates a contingent ideal of life as a technique of mastering pressing disorders (Stoler 2008: 350). In a compounding history of usage, defending the environment has been sharpened into a kind of operational common sense that valorizes the quality of ordinary life after the fact of its disruption, and by so doing provides a practical means for both objectifying that disruption and orienting restoration. Following Canguilhem ([1966] 1991), it is the unexpected rupture of a disaster that keys us in to what we should have known all along and now must effectively conjure up in its absence: namely, the science of the normal. I take the environment to be just such an ex post facto science of the normal.

PERFORMING CERTAINTY

Arriving well before dawn was the key to securing a seat, and even then there was already a crowd. Congressional hearings often only have space for about fifty members of the public. But when a high-profile hearing occurs, those seats can get whittled down to a handful as media and congressional staffers claim the public seats. Spilling out of vans, a dozen college students gathered on the sidewalk outside the Rayburn Congressional Building, half awake as they put on new T-shirts emblazoned in block letters: "OIL SHOULDN'T COST LIVES" and "CHALLENGE CORPORATE POWER." Coffee brought the stragglers into a huddle as they worked out a homegrown strategy to secure front row seats. I struck up a conversation with the leader, saying it must be hard to get twenty year-olds to get up before 5:00 a.m. for a congressional hearing. He laughed. They don't really have a choice, he said; as interns for Greenpeace, attending the hearing is required. "We teach them what it takes to be an environmental activist," he said. On the other side of the main entrance, a number of diverse men and women in dark business suits sat on cheap lawn chairs. Calling themselves "professional line workers," they explained their job. They had responded to a craigslist ad looking for "diverse professionals." When a high-profile hearing came up, they were paid handsomely to dress conservatively and occupy the seats most likely to be included in any media coverage of the proceedings. Although things had been quiet in this industry for some time, there was rising excitement about the prospects of the coming BP hearings. To prevent the televised image of righteous outrage framing the hearing, corporate lobbyists hire people of color to occupy public seats in solemn seriousness.

Security opened the doors to the building at 6:00 a.m., and both sides scrambled through in a rush. BP's CEO Tony Hayward was set to testify. I was the first through security and arrived at the committee room in high hopes of securing a seat, only to find three activists and three line workers already there (two congressmen had snuck them in early, they indicated). Behind me, the line soon stretched into the hundreds. (One sound guy told me he hadn't seen this many people trying to get into a congressional hearing since Frank Zappa was hauled in to shed light on profanity in rock music in 1985.) Standing at the head of the line, three women were dressed as hot pink seals. They were Code Pink, an environmental advocacy group. (I explained my research interests to one of them. She scoffed, "It's past time for studying the oil industry; it's time to disrupt it.") Seats were scarce. After making room for the media, there were only four seats for the public. Fortuitously, at least for me, the Code Pink activists dumped crude oil on themselves, just as Tony Hayward was being seated, and were hauled out by the police. After the three suits were seated, I got the last public spot (a seat slightly stained with crude oil), squeezed in between a *Wall Street Journal* reporter and a *Financial Times* reporter and exactly one row behind the main event: Tony Hayward.

On the surface of soundbites and stagecraft, the hearing was rather uneventful. Members of the committee lambasted BP for its inability to rein in the oil spill or even properly explain the size of it to Congress. Tony Hayward expressed his deep regret for "the incident." Although Tony Hayward's legs trembled underneath the table, everything above the table was a perfect image of calm and contrition. He slowly enunciated his many apologies while stopping just short of admitting any responsibility, couching his responses in careful qualifications like "We do not know at this time" and "I am sincerely sorry but we do not have an answer to that question at this time." Most of the allotted time was taken up by the members of the committee themselves, whose questions had a way of slipping into grandstanding. Many spoke not to Tony

Hayward but toward the cameras in the back of the room. At one point, all the photojournalists in the room got up and rushed to a side door in a flurry of clicks and flashes as a man stepped in to watch the proceedings for a few minutes. It was Kevin Costner.

Most national papers had a front-page story of the proceedings published online before the hearing was even over. I watched several of the stories being written, most notably by the reporters on both sides of me. Both journalists were working on oversized laptops that, in the cramped conditions of the committee room, rested on my lap. Ignoring the hearing, the *Financial Times* reporter played and replayed a video of Tony Hayward entering the room as he tried to find the right words to describe the scene. The *Wall Street Journal* reporter was emailing his article, paragraph by paragraph, to someone who sent back substantial revisions a few minutes later. It took me a few minutes to realize that "the editor" was in fact sitting in front of me and was no other than BP's chief counsel overseeing the oil spill. I watched her delete and insert lines on her Blackberry, softening words here and there and watering down several key passages about BP's potential liability. She had given the *Wall Street Journal* a scoop, I found out, by leaking key details of President Barack Obama's dressing down of Tony Hayward in the previous day's private meeting at the White House. She offered an inside perspective on the company under the alias "a high-ranking official with BP." The price, apparently, was that she personally edited the resulting coverage. After haggling for the better part of an hour, the *Wall Street Journal* reporter emailed his article to his newsroom editor. It was posted on their website a few minutes later and ran the following day in the print edition.

My interest in this "high-ranking official with BP" was piqued, and I took note. I took note when she wrote a colleague to calm fears; she had secured a hand in shaping the reporting of a handful of national newspapers and regional ones along the Gulf Coast in Texas, Louisiana, Mississippi, and Alabama. I took note when she directed another colleague to hire all of the drilling concrete experts they could find and do it quickly (this was a few days before the issue of BP's having directed Halliburton to use a substandard concrete mix in the Macondo well was disclosed). And I took note when she started firing off emails asking about university-based marine scientists who were starting to study the oil spill.

Some of those marine scientists later explained what happened next. BP contacted them, offering them nearly unlimited research resources, from funds for post-docs to updated laboratory equipment to research vessels to unfettered access to the wellhead. "I could demand any test," one told me, "but there was gag order on my data. I could not publish any of my data." The resources went beyond research. BP also offered to pay some of these scientists over $100,000 a year to consult on the oil spill. This offer came from BP's legal team, which effectively removed all of their publications and research from the public record and encased it within attorney-client privilege. As was reported by *Mobile Press-Register*, this offer was extended to entire marine science departments at several Gulf Coast universities (Raines 2010). After being approached by BP, the head of the Dauphin Island Sea Lab commented, "They were more interested in making sure we couldn't testify against them than in having us testify for them" (Raines 2010). The unparalleled ability to study the oil spill came with the proviso that these scientists' entire body of scholarship would be effectively insulated from any public relevance. Richard Shaw, dean at LSU's School of the Coast and the Environment, noted that these lucrative opportunities intersected with years of declining federal and state funding for marine science research along the Gulf. Shaw said, "People are signing on with BP because the government funding to the universities has been so limited. It's a sad state of affairs" (Raines 2010).

The newfound significance of the deepwater made its way into the congressional hearing as well. Throughout the proceedings, a number of committee members asked Tony Hayward about early reports of "underwater plumes of oil" that were, at this point, starting to come to light in academic blogposts and regional newspapers. "We have no evidence of that," Hayward replied flatly each time. He stressed that BP was spending a considerable amount of money funding independent scientists to try to replicate the results of these so-called underwater plumes, so far to no avail. They simply could not find them. And Hayward pointed out that the National Oceanic and Atmospheric Administration (NOAA) had yet to independently verify any of the reports of underwater plumes. As BP's CEO testified that the company had no evidence of underwater plumes, its top lawyer—a mere arm's length away—was purchasing all the expertise that might challenge such a brazen lie.

In these congressional hearings, the performed certainties of preexisting positions overshadowed the unanswered fundamentals of the spill. Nothing seemed to reach beyond fixed partisan stances; nothing seemed to grapple with how uncontrollable this deepwater blowout was proving to be. I wondered where these unprecedented properties of the oil spill were actually present. The next week I drove along the Gulf Coast and met a number of bar owners, shrimpers, cleanup workers, small town doctors, marine scientists, and other coastal residents beset by the uncertainty of the ongoing oil spill. So much of their questions and anxieties centered on the ecological reach of the oil spill and the sense that the oil was already among them. Talking with them, I realized that foregrounding this uncertainty might also provide a better ethnographic approach to the disaster at hand. Many social scientists studying disasters arrive just after the event, when the disruption already has a certain public definition that experiences can then be understood in relation to. Here, in the midst of an unfolding and unprecedentedly large oil spill, the experience of disruption preceded and exceeded any clear definition of disaster. I realized that if I wanted to understand the materiality of the oil spill, I would need to figure out where these unknowns were an active problem. Whether in research universities or in government agencies, marine science was taking up the deepwater blowout as a new kind of scientific problem. But what kind of scientific problem is an oil spill?

THE ENVIRONMENT MUST BE DEFENDED

On April 27, 2010, the still uncertain aftermath of the Deepwater Horizon's sinking was designated an "Oil Spill of National Significance" by Secretary of Homeland Security Janet Napolitano. That formal designation mobilizes the Unified Command System and authorizes it to take control of the situation. The Unified Command System, at its peak, came to employ "more than 47,000 personnel; 7,000 vessels; 120 aircraft; and the participation of scores of federal, state, and local agencies" (Mabus 2010: 2). This infrastructure of response resembles an event-centered government agency, an interdisciplinary department whose authority (and lifespan) is tied to solving an urgent problem. Unified Command was given exceptional authority to, as I was told repeatedly, "protect the public and protect the environment." These mandates were initially taken to be two largely overlapping operations. Both consisted of keeping crude oil off the coast. However, as the deepwater dimensions of this oil spill slowly came into focus, the protection of the environment became a separate problem that moved away from the populated coast and into the alien world of the deepwater.

The Unified Command System gained its current authority from the Oil Pollution Act of 1990 (OPA). Coming on the heels of the *Exxon Valdez* spill, the Oil Pollution Act of 1990 (OPA) legislated double-hulled tankers, navigational aid for major energy ports, and a more robust means of assigning liability. Indeed, much of the language in OPA reflects the particular setting of the *Exxon Valdez* spill and its iconic devastation of the rugged coastline of Prince William Sound. The ocean, for example, is treated as a corridor that can transport crude oil into sensitive sites like beaches or spawning grounds. The ocean is treated not as a dynamic ecological system in itself but as a pathway to human and animal exposure. In other words, OPA projected a leaking supertanker as the definitive oil spill and animals and beaches coated in crude as emblematic of the environment in need of protection.

Environmental protections in the United States often formalize the historical contingencies of a single disruptive event as the generic conditions of all future disasters. Such risk management, as Charles Perrow (1984), Lynn Eden (2004), Michael Powers (2004), Stephen Collier (2008), and Limor Samimian-Darash (2009) have demonstrated in different contexts, works to reduce the shifting complexities of the present to certain reified forms, a process that removes questions of temporal and spatial specificity and deploys a now standardized political calculus of technocratic risk (Clarke 2005). To prevent another *Exxon Valdez*, the OPA directed federal agencies to join together and prepare for the worst oil spill imaginable. In subsequent statutes and ongoing disaster preparations, *Exxon Valdez* became the de facto worst-case scenario that the Unified Command System calibrated itself to remedy. The last disaster became the new governing norm.

During the first month of the BP oil spill, the protection of the environment largely consisted of a mechanical application of historical insights: namely, what should have been done during the *Exxon Valdez* spill. Tens of thousands of miles of boom were ordered to line the coast from Texas to Florida, while chemical dispersants were readily approved not only for surface application but also for deepwater injection (for which there was little precedent). The logic behind these strategies was the same: protecting the environment was all about preventing crude oil from making landfall. Federal agencies, in other words, sought to protect the environment they already knew how to protect. The orientation of their operations took its cues from historical precedent. "What we apply to the next spill is what we learned from the previous spill," the lead environmental scientist explained to me at the beginning of the BP oil spill.[4]

"Most of the technology we are using now was developed in Exxon Valdez," NOAA scientist Charles Henry told me early on in the BP oil spill. Henry had worked in Alaska for the decade following the *Exxon Valdez* disaster and headed up the environmental science division within Unified Command. The legislative and technological mastery over the problems of the *Exxon Valdez* spill was apparent in the emergency response to the BP oil spill. Unified Command bustled with *Exxon Valdez* veterans, proudly brandishing their experience in Alaska at press conferences and in planning meetings. The point was clear: they had been here before and they knew what to do. There was a confidence at Unified Command in the early days of the BP oil spill that suggested oil spills were all the same. They were generic events that, at least since *Exxon Valdez*, had been scientifically deciphered. And now any oil spill could be mastered with the right application of technology and operational expertise. Indeed, with advanced degrees in oceanography and experience with the cleanup of the *Exxon Valdez* spill, many of the personnel of Unified Command described themselves to me as experts in "the science of oil spills."

MAY 2010: THE DEEPWATER

On May 6, University of Mississippi marine scientists Vernon Asper and Arne Diercks directed the research vessel *Pelican* toward the Deepwater Horizon site. It was a hastily organized and then reorganized research trip on which they intended to take mud samples from the Mississippi River Delta and keep a formal log of any surface sheens of crude oil they encountered. The two marine biologists were continually taken aback by the lack of crude oil on the ocean's surface. "I really expected the oil to be on the surface," Asper told me later. Despite their vigilance, they could not find any oil on the surface. So they started looking for it elsewhere. According to their blog, they tweaked and repurposed several monitoring technologies to try to see what was happening under the surface. On May 12, their underwater sampling devices picked up "the presence of several layers of 'material' at depths from 700 to over 1300m" (Asper 2010: 2). After additional testing, they hypothesized that these layers of "material" might in fact be violently emulsified (and chemically diluted) crude oil from the Deepwater Horizon that somehow had remained suspended in the water columns. These questions spread through emails to colleagues and datasets shared with media outlets. Within a few weeks, multiple university research vessels (R/Vs) were testing for the presence of suspended hydrocarbons in the deep sea. Although these marine scientists had no experience with oil spills, they possessed the technical know-how to see what was happening under the surface. Quickly ruling out alternative sources of the oil, the team pinpointed the spewing wellhead as the author of these underwater plumes. "We've taken molecular isotopic approaches," one researcher told CNN, "Which is like a fingerprint on a smoking gun" (Couwels 2010: 2).

Such findings were initially dismissed by BP. "The oil is on the surface," Tony Hayward asserted. "There aren't any plumes"(cited in Couwels 2010: 2). BP contracted its own research vessels, all of which tested for submerged oil using traditional metrics and concluded, "There is no coherent body of hydrocarbons beneath the surface"(BP 2010c: 1). BP's own cadre of scientists rejected the proposition with common sense physics: oil is lighter than water and thus must rise to the surface (BP 2010a, 2010b). The implications were clear: all efforts to measure and contain the ongoing oil spill should focus on the ocean's surface (BP 2010a, 2010d). As the *New York Times* reported: "BP has resisted entreaties from scientists that they be allowed to use sophisticated instruments at the ocean floor that would give a far more accurate picture of how much oil is really gushing from the well. 'The answer is no to that,' a BP spokesman, Tom Mueller, said. [. . ...] 'It's not relevant to the response effort, and it might even detract from the response effort'" (Gillis 2010a).

The federal response to this debate was initially quite tepid. NOAA administrator Jane Lubchenco held a press conference after the R/V *Pelican* began releasing its data: "We do not yet have a good handle on the question of subsurface oil" (cited in Gillis 2010b: 2). A NOAA report on the matter stated: "No definitive conclusions have been reached by this research team about the composition of the undersea layers they discovered" (2010: 1). "The business of trying to detect submerged oil is not a settled science," a NOAA official told a reporter (cited in Clark 2010: 1), a sentiment later repeated to me almost verbatim. NOAA tried to independently verify the findings of the *Pelican* and other university R/Vs by analyzing their deepwater samples in NOAA's lab. Although hydrocarbons were present in all the samples, concentrations were below the "eco-toxicology benchmark" that defines a legally reprehensible oil spill (Haddad and Muraski 2010: 2). They were below the threshold that bounded an official definition of an oil spill. Indeed, the hydrocarbons were so diluted in the water samples that it was impossible to chemically identify their source. With existing technology, the hydrocarbon "concentrations are too low to allow for source determination"

FIGURES 4 and 5. (Left) Glass flask containing discrete oil, Joye Lab, University of Georgia–Athens. (Right) Glass flask containing dissolved oil, Joye Lab, University of Georgia–Athens Photos by author.

(Haddad and Muraski 2010: 2). If the matter of origin could not be established with some degree of certainty (and with it liability), then the scientific and technical resources of federal agencies were best deployed elsewhere.

JUNE 2010: THE SCIENCES OF AN OIL SPILL

The BP oil spill became the recipient of urgent scientific questions without first being stabilized as a clearly defined scientific object. Perhaps this is always the case with disaster; the pressing task of knowing how bad it is precedes any agreement on what counts as valid data. In disasters, as in the history of science itself, analytical practices unfold beyond the clearly defined norms of a scientific community (Shapin 1994). The deepwater, a very acrimonious and publicized topic, was at the very crux of this uncertainty during the BP oil spill. "Underwater plumes" were headline news. As reported, this was a standoff of sorts between political bureaucrats and academic scientists, and the story often leaned toward a tale of government incompetence. Up close, this debate was something else entirely. Media coverage skirted two details. One, this was not politics versus science but a debate between two groups of marine scientists, one affiliated with federal agencies, the other with research universities. Two, the content of this debate came to rest on an exceedingly practical question: Which subsea device was best suited to monitoring dispersed hydrocarbons in the deepwater?

During the second and third months of the BP oil spill, academic scientists like Vernon Asper and Arne Diercks working in the Gulf independently of Unified Command began to notice cascading changes in the ocean itself. These marine scientists began with a precise but limited sense of various niches or species within the ocean; as they noticed specific disruptions in their field of expertise, they began rethinking the dimensions of the oil spill and the environment it imperiled. Fluent in the interrelatedness of the ocean, those specializing in the biogeochemistry of the deepwater began working out a plausible link between the oil spill, microbes, and observed shifts in the chemistry of the deepwater. Microbes consume hydrocarbons by drawing them into a chemical reaction with oxygen. This biologically mediated reaction consumes hydrocarbons and oxygen and produces carbon dioxide. As they break down oil, microbes alter the chemistry of the water in durable ways. Focusing in on the altered chemical and microbial state of the deepwater, marine scientists devised their own means of discerning crude oil. Collating the effects, they redefined the cause. These emerging definitions, while illuminating the specific properties of diffuse hydrocarbons in this spill (and its shockingly large scale), were at odds with Unified Command's more established means of locating and remediating crude oil in the ocean. Indeed, the technical ability to remediate often provided an informal map that guided where Unified Command searched for the oil spill. In the habitus of the emergency response, if spilled oil was too diffuse to capture then it was of secondary concern. What mattered most was oil you could do something about.

Although they shared degrees and a basic analytical language of the ocean, marine scientists working in research universities and marine scientists working in federal agencies differed on how they produced new knowledge about the BP oil spill. Academic scientists sought to produce *disciplinary facts*, focusing attention on documenting the specific impact of dispersed hydrocarbons on the ecological niche or species they knew best. Government scientists, in contrast, sought to produce *operational facts* that could help rein in an unfolding event, focusing attention on proven technologies of surface measurement and capture of an oil spill. This difference was widely discussed and palpable in operational meetings throughout the spill. Academic scientists often described themselves as belonging to a "research community," while government scientists emphasized their commitment to solving real problems with science. Although both groups of scientists positioned the significance of their work in relationship to the environment imperiled by the BP oil spill, the respective venues of their research shaped how they initially saw the oil spill.

JULY 2010: THE TECHNOLOGY OF CONSENSUS

As the BP oil spill continued unabated, one official told me: "We knew how to respond to a surface spill, but this is completely different." As one scientist working with NOAA summarized: "Exxon Valdez released oil on the surface for about 12 hours. This spill has been going on a mile underwater for nearly 3 months now." The limits of the official science of hydrocarbon disasters soon became apparent. "The oil spill regulations written after Exxon Valdez were written for the next Exxon Valdez," Admiral Thad Allen, who headed up Unified Command, told a group of congressional aides and federal regulators in Washington, D.C., a few months after the BP oil spill. "In retrospect, OPA is one hell of a tanker-centric piece of legislation." The oil spill he dealt with, he explained, was a deepwater blowout that was "indeterminate and multidirectional, it was disaggregated and going in different directions." He summarized: "We could barely keep up with it." The effects of the BP oil spill, both at the underwater site of the blowout and in the immense scale of the spill itself, soon surpassed the legislated and practiced norms of the defendable environment.

It became clear that some agreement between marine scientists working in universities and marine scientists working for regulatory agencies— a new collaboration between disciplinary facts and operational facts— would be the only way forward. This debate between these two venues of science, played out in conference rooms during the oil spill, eventually led to a working consensus on which technologies to use to measure the deepwater dimensions of this oil spill. On its face, this consensus was premised on the growing need to establish a standard metric for detecting subsea oil that could be applied uniformly across the entire Gulf of Mexico. Yet something else occurred as well. Agreement on the key technologies to measure the disaster also rendered the deepwater as a stable (and now sequestered) field of calculation for those devices (Pinch and Bijker 1984; Callon 1989). The practical metrics of the chosen technology mapped out the deepwater of the Gulf of Mexico as a static grid of vulnerability within which disruptions could then be scientifically documented and validated; these technologies produced the deepwater as an environment now formatted for thresholds and impact assessments.

At least initially, scientists working with Unified Command were reluctant to monitor the impact of dispersed oil in the deepwater if nothing could be done about it. Their mandate was to protect the environment; research that could not "yield real-time operational results," as one directive put it (USCG 2010), had no place in the emergency response (and the resources of that exceptional authority). "We are not assessing the long-term impacts to the environment," one scientist working with Unified Command told me as the oil spill stretched into its third month. "We are addressing immediate threats to the environment that we can mitigate and amend." "Everything we do in an emergency response," another government scientist said, is tied to "decision points" or "actionable levels." I overheard one federal scientist paraphrase it this way: "We can only act if we can do something about the threat."

Right "now we are looking for oil that can be remediated," one government scientist told a group of academic scientists at another meeting during the oil spill. "The other discussion is the ecological impact of the oil," he added. "That discussion will happen later." A bit later an exasperated official responded to continued criticism of the lack of attention directed to dispersed oil in this way: "There is nothing we can do about deepwater plumes. We can't pump that oil out, that's a mile deep!" For the first few months of the BP oil spill the mandate to produce useful knowledge led government scientists (and federal officials) to actively exclude microbial and chemical evidence of deepwater plumes of hydrocarbons. Microbiology in the deepwater was not part of the defendable environment. As evidence of that alien disruption grew, however, the form of the defendable environment was destabilized.

Marine scientists working in research universities first discovered dispersed hydrocarbons in the ocean by creatively repurposing a fluorometer. When I asked one academic scientist if the fluorometer was intended to work in oily water, he laughed, "None of our equipment was designed to work in an oil spill." Another told me, "Oceanography isn't supposed to have oil in it." The fluorometer was originally designed to provide instant monitoring of the organic (i.e., carbon) composition of water, which indicated the presence of plankton or algae by emitting fluorescent light and monitoring the colors reflected by the water. During the BP oil spill it was retrofitted and recalibrated to indicate the presence of dissolved or dispersed hydrocarbons in the deepwater. The improvised use of the fluorometer by academic scientists quickly became the basic technology through which the underwater dispersal of hydrocarbons could be seen in the ocean's depths (and sampled for laboratory analysis). Although few if any ever intended to study a deepwater blowout, marine scientists in research universities possessed the technology to see the basic medium of just such a disaster: namely, hydrocarbon-saturated seawater.

Reflecting on these innovations, one marine scientist talked about how exciting this new "interdisciplinary research" was. "All the major experts are on board," he said. "We have unfolding and flexible research plans that can adjust to preliminary results. The florometer can find the oil. Biologists can see what it's doing. Chemists can figure out where it's from." Running simultaneous tests and adjusting the research plan to the results, he said, has led to a "broadening of techniques" to understand the deepwater environment. "Academics and industry have broadened their techniques. Unified Command needs to catch up." Academic scientists were quite clear about what this technology and the insights it enabled meant. The BP oil spill, oceanographer David Hollander told CNN (August 17, 2010), has "changed the paradigm of what an oil spill is from a two-dimensional surface disaster to a three-dimensional catastrophe." "Through this all, we have witnessed an aged and untested bit of dogma dominate response decisions: Protect the beach," oceanographer Robert Carney told National Geographic (August 19, 2010). "Quite obviously, it is the whole ocean that we must protect and effectively manage." Although there is a "gut reaction" to crude oil on the surface, marine scientist Samantha Joye told me, "I think when all the data is in, the subsurface effects are going to be far, far more extensive and far, far more long-term than the effects of the oil that made it to the surface."

As it became clear that most of the crude oil in this spill neither made it to the surface nor threatened coastal areas—a NOAA report in September 2010 found that less than 10 percent of the spilled oil was recoverable at the surface—the federal mandate to protect the environment was tentatively shifted from coastal protections to figuring out where all the oil went (NOAA 2010). In late August, Unified Command released a subsurface "detection, sampling, and monitoring strategy." Although this directive aimed to produce actionable data, it chronologically and operationally separated the question of where the oil went from the question of what could be done about it. "Monitor and assess the distribution, concentration, and degradation of the remaining portion of the oil that remains in the water column and/or bottom sediments" became an operation in and of itself, one that preceded the subsequent task: "identify any additional response requirements that may be necessary to address remaining sub-surface oil." Mapping the scope and subsurface movements of the spilled oil became part of the emergency response, even if nothing could be immediately done about it. This, it bears pointing out, was an unprecedented operation. The deepwater is an extreme environment; it cannot be surveyed in its entirety but rather must be sampled at precise points and general conclusions drawn from those samples (Helmreich 2009).

Unified Command set up a series of meetings between academic scientists who were researching the deepwater movements and effects of the oil spill and scientists working in federal agencies responding to the spill. The goal of these meetings was to establish sampling standards for locating and measuring dispersed hydrocarbons in the deepwater. At one meeting, a government scientist vented his frustration: "How far should we track and map the oil? Part per million? Parts per billion? Part per gazillion? You get to a point where you are chasing things that aren't real." The academic scientist countered: "Dispersed oil is still toxic, we're seeing real toxicity." At another meeting, an academic scientist asked, "Toxicity levels for tropical species might be lower than our capacity to measure. Can we detect what we need to detect?" When I ask Steve Murawski, a NOAA scientist working with Unified Command, about new efforts to monitor hydrocarbons in the deepwater, he told me: "It's a mile underwater. It's extremely hard to characterize what's going on down there. And Unified Command wants to know for certain." That certainty came to rest on the limits of technology that could be widely and uniformly applied. This was not a matter of how fine a scale or how far away the effects of oil could be seen; it was the far more practical question of how to monitor and measure the deepwater in a standardized manner. As Lorraine Daston and Peter Galison (2007) might suggest, it is the agreed upon way of looking into a disaster that transforms it into something amenable to objective knowledge.

AUGUST 2010: PLACING THE PROBLEM

In the weeks after the wellhead was capped, the emergency response began focusing its resources on the technical stabilization of the deepwater. "This oil spill is maturing a lot of technology," Steve Murawski said at the last meeting between academic and government scientists. He referred to the fluorometer but also played up the increased usage of remote sensors and underwater gliders. These new technologies of automated surveillance, he noted, "are a whole lot cheaper than taking a boat out and splashing water." And unlike fluorometers, they can be calibrated to a common standard.[5] "There's been a real economy of technologies here," he said, describing the way the improvised use of the fluorometer had cracked open the door to a new dimension of environmental impact and fostered the development of more rigorous monitoring and measuring technologies. "We've had to change technologies throughout this process," he continued, and now it was time to settle on the best available technology and put it to work on an immense scale. He laid out new plans to place remote sensors throughout the deepwater on a grid organized around the wellhead. Subparts per billion of select hydrocarbons would be the effective threshold.[6] We needed "a statistically valid sampling plan" for the entire Gulf of Mexico, he said, and we needed "to format the data so it is compatible" with all other research operations working to "determine where the oil is in the environment." "We will sample until we have a good representation of where the oil is." The goal, he concluded, was to produce commensurable data on the full width and breadth and depth of this oil spill that could be housed within a single database. "Putting a complete picture together is key," he said.

During the BP oil spill, the affected ocean was, in a way, transformed into a scientific laboratory within which the true size and scope of the oil spill could finally be mastered. This "laboratorization" of the Gulf of Mexico (Callon, Lascoumes, and Barthe 2009: 65), again, had less to do with documenting the outer reaches of hydrocarbon effects than with the technological monopolization of method and its implicit consolidation of hydrocarbon facts (Latour and Woolgar [1979] 1986; Pinch and Bijker 1984). The materiality of the oil spill was redrawn around the analytical capacity of select devices. This technological consensus transformed the varied scientific inquiries that gathered around the oil spill into a "science for policy" of the oil spill (der Sluijs et al 1998: 315). That is, agreement on how to measure the oil spill was also an agreement to understand the oil spill in a way that leaned toward the pragmatics of state management (Jasanoff 1990; Lahsen 2009). Scientists who wanted their work to be relevant to the emergency response (and the enormous resources it offered) had to discipline their questions and findings into the technical configuration of facts deemed legitimate by the state.

Disasters, as Kenneth Hewitt (1983: 10) argued some thirty years ago, "are made manageable by an extreme narrowing of the range of interpretations and acceptable evidence." The environment, first overwhelmed by the fractured quality of this deepwater blowout, was adjusted to contain this multivalent disaster as an unequivocally singular event. Reworking the baseline conditions of the ocean around the background detection capacity of select subsea devices, the environment was expressed as a standardized grid of subsea chemical conditions against which the BP oil spill could finally be seen as a discrete disruption. It is, as Hewitt (10) put it, this "careful, pragmatic, and disarming *placement* of the problem" that comes to "fix" disaster in both senses of the word: it bounds the disruption in time and space and orients recovery. Such fixing also produced a new boundary between the oil spill and everything else. The BP oil spill changed from a sprawling mess into a manageable problem by being lodged within a refined deployment of the environment. This placement of the problem, as Tim Forsyth and Andrew Walker (2008: 233) have described elsewhere, provides "a seemingly neutral justification for selective state action." It was, to borrow a phrase from Theodore Porter (1995: 8), "a way of making decisions without seeming to decide."

SCIENCE IN THE WAKE

Afterward, after the wellhead was plugged, after the deepwater environment was formalized, after the spill was measured and mapped, and after the experiences of residents were rendered illegible and irrational, the long work of restoration began. This process brought the texture and trajectory of yet another fact into the work of science around the oil spill: legal facts. The ecological investigation had begun. The task, formalized in the Natural Resource Damage Assessment (NRDA) process, constructed a set of procedures to give the destructive reach of an oil spill a fixed value that might be equally persuasive in scientific laboratories, courtroom proceedings, and restoration projects. While still framed as an open investigation conducted in the public's interest, in practice "damage assessment" has become attentive to crafting facts that anticipate future legal disputes. As one federal official described the ratcheting up of proof in NRDA to me: "You can have good science, you can have great science, but it won't hold up in NRDA. You have to have the absolute very best science. Anything else will get tossed out." Another one summarized the work of NRDA in this way: "We use science to build a case." In practice, what has come to matter for NRDA is less the documentable forms of life impacted by an oil spill and more the kind of facts that might withstand multiple forms of scrutiny. While this may be a perfectly reasonable strategy, it does come with a cost. As "damage assessments" have come to anticipate and in some cases internalize the very corporate strategies their work is designed to confront, public voices and concerns have been quietly pushed to the side.

The anthropology of science has taught us that, despite appearances to the contrary, the texture of a scientific fact is always embroiled in wider disputes. There was no such appearance to the scientific production of these facts in the wake of the oil spill; it was not the absence of interest that certified their credibility but the abundance of interest that imbued them with agreed upon stability. After the *Exxon Valdez* spill and in consultation with local communities, federal scientists produced the government's own dataset of the disaster, carefully locating, measuring, and aggregating injuries to natural resources. It was an impressive, wide-ranging account of the disaster. However, when that dataset entered the courtroom, it was confronted with Exxon Corporation's own dataset. The subsequent twenty years of legal wrangling have largely been a debate over who has the best science of petro-destruction, the state or the corporation. Determined to avoid such quibbling the next time around, federal agencies now team up with the guilty party in an oil spill to produce a single dataset of damage to natural resources. As one NRDA official told me during the BP spill: "You can argue a lot on the meaning of the data, but at least now there's only one dataset."

During and after the BP oil spill, teams of federal scientists and BP representatives worked hand in hand to establish a single "damage assessment" of impacted natural resources. Every injury had to be verified and agreed upon by both parties. Reportedly, most NRDA research trips were staffed with one NOAA scientist and one attorney representing BP. During the spill, I heard numerous stories of oiled birds and bloated fish that never made it into the official record for lack of a BP lawyer to witness their morbidity. The significance of this strategic cooperation, however, goes beyond abridging the official account. As negotiations between state agencies and the guilty corporation over the science of destruction are now hidden from view, their contingent compromises are presented not as the outcome of an ongoing conflict but as fully formed and state-endorsed fact. Such concerted fact production occludes the experience of local residents from being an active register of the oil spill. And yet, simultaneous to that technopolitical disregard, such residents are gathered as witnesses needed to legitimate this off-stage science and authorize the interventions it foreshadows.

Attention to this dispute challenges some popular approaches to toxic exposures, namely that the present plight of many fence-line communities might be resolved by simply adding science to the mix. In such approaches, science is held up as a radically disinterested vehicle that might finally chart a clear path through the entrenched interests of the state and the corporation. Such science is certainly helpful and often very much needed. But such claims can overlook the tremendous investments the state and corporations have already made in the science of hydrocarbon destruction. Perhaps the cultivated disregard for local residents in such sites is the result not of a lack of science but of a sort of scientific arms race between state agencies and harmful industries.

THE NEW NORMAL

In the aftermath of the Santa Barbara oil spill of 1969, sociologist Harvey Molotch suggested that disasters are opportune moments for scholars because they expose the underlying social relations that in ordinary times would be blurred or inaccessible. "This technological 'accident,' like all accidents," Molotch (1970: 131) writes, "provides clues to the realities of social structure (in this instance, power arrangements) not otherwise available to the outside observer." Disasters, William Torry (1979: 517) summarized in a review article in *Current Anthropology*, "draw into sharp relief a variety of fundamental processes less easy to observe or interpret in more ordinary times." This insight has been amplified in a growing body of scholarship in science and technology studies (STS) that suggests disasters are "normal events" insofar as they reveal the otherwise ignored embeddedness of technological risks within the social (Perrow 1984; Jasanoff 1984; Vaughan 1997; see also the February 2007 issue of *Social Studies of Science*). Disasters are technopolitical exposés of the highest order. In this chapter, I have argued in a different direction. Destruction carries its own rippling creativity. While disasters may reveal entanglements we long suspected, they also work to resignify and order the impacted world anew (Erikson 1976; Das 1995; Vaughn 2012). Disasters, then, do not only reflect an "extreme version of everyday life," as Edward Woodhouse (2011: 61) put it, they also instantiate new knowledge of life.

"We know terribly little about what the deepwater was like before the spill," one marine scientist told me. The chair of the national commission investigating the oil spill has said as much: "One thing we learned is how little we know about the basic environment in which the crisis took place." (Or, as a lab technician testing water samples from the oil spill explained to me, "I love science, but this is one fucked up science experiment. There is no control.") In fact, much of what we now know about the deepwater is, in many ways, a direct result of the BP oil spill. "In the last 3 to 4 months there has been an upsurge in knowledge about the Gulf. We understand the Gulf better now than we ever have," one NOAA official told me. For the first time, there is uniform data on the microbial and chemical composition of the deepwater across large swathes of the Gulf.

This emerging "environment" of the BP oil spill is fast becoming an immensely productive field for new forms of scientific inquiry and political responsibility. "Our toolkit has evolved tremendously in this spill," one Unified Command official told university officials several months after the wellhead was capped. "The bottom line is we need to learn from this one, we need new knowledge," he said, announcing a $500 million research initiative in the Gulf of Mexico to study the environmental impact of the spill.[7] More recently, the US Department of Justice announced that nearly half of its criminal settlement with BP would be earmarked for environmental projects in the Gulf. In response to this "unprecedented environmental catastrophe," US attorney general Eric Holder explained that $2.4 billion would be "dedicated to environmental restoration, preservation, and conservation efforts" in the impacted region (2012). This burst of funding, attention, and data, I suggest, is less a definitive accounting of ordinary biology in the Gulf of Mexico than a persuasive mapping out of a new domain of calculation and administration. It is a scientific instantiation of the new normal.

AFTER THE FACT

Disasters are productive events. Recently, popular (Klein 2007) and scholarly (Gunewardena and Schuller 2008; Lakoff 2010) attention alike has focused on the ways disasters can open the door to neoliberal restructuring. Less attention, perhaps, has been paid to the epistemic urgency of disasters; that is, how disasters demand definition and the social consequences of how they are defined. Following Canguilhem, I have shown how the official response to disasters like the BP oil spill cultivates an emergent understanding of normal life.[8] This instigated normality both works to define the extent of the disaster and becomes the foundation for a new regime of thresholds and impact assessments. This instigated normality also offers itself as a platform of sorts for subsequent scientific, political, and ethical projects (without, in either case, becoming an object of much scrutiny).[9] In more ways than one, the last disaster becomes the new governing norm.

The environment—the knowable and governable conditions of life—is not some staid figure but rather an unruly process continually given new delineations and new momentum by unexpected disruptions, like disasters (or the threat of disasters; see Masco 2010). During the BP oil spill, the environment came apart and was put back together again as the constitutive normal that reined in the disaster. Staying close to the embedded operations of the state during the oil spill, the emergent environment described here is neither a culturally interpellated place (Rappaport 1968; Balée 1994; Raffles 2002) nor a discrete fenced-off place (Walley 2004; Lowe 2006; West 2006; Welker 2009) so much as it is a material effect of an expedient assemblage of sampling devices and their detection capacity. As I have shown, these devices articulated a working definition of the baseline conditions of life that both objectified the disaster and oriented scientific practices within (and after) the disaster. The critical question of the constitution of normality, then, is not always one of the intentional impositions of power but also one of the distribution and density of monitoring technologies. Quietly orienting the state's forceful considerations as well as its averted gazes, these sovereign networks of sampling devices enliven the governable environment (Allen 2003; Fortun 2012). Within such networks—far more proprietary than emancipatory—disaster (or even danger) is depicted not as a risky calculation tangled up in industrial investments and demographic expediency but as a disembodied scientific object measured against an implemented baseline.

Almost a year after the BP oil spill, I traveled to DC for a meeting among federal agencies to reflect on how threats to the environment were addressed during the spill. In a marbled hallway afterward, I ended up at the cookie table with the senior official who had headed up efforts to protect the environment during the spill. He explained that although this deepwater blowout initially overwhelmed the emergency response efforts, its impact was eventually brought into sharp focus with the right science. "This is the problem of science. We put together a model, find the limits of that model, and then build a better model," he said. "As bad as this oil spill was, it's been great for science."

CHAPTER 3

Ethical Oil

One of the more remarkable things about the tar sands in Alberta is how upfront oil companies are about their impact on the landscape. On local billboards and in interviews, tar sands operators regularly acknowledge that they are going to destroy the place. After all, they say, this is "the real cost of energy today." Such acknowledgments, however, quickly pivot toward the huge investments the industry is making in their ability to put it all back together again. With restoration projects that strive to join the best of environmental science and the most traditional Indigenous ways of life, oil companies proudly tout their unique ability to engineer a more culturally informed boreal forest, to build a more cosmologically attuned northern ecosystem. With input from cultural consulting firms, Steve, the director of aboriginal relations at a major tar sand project, told me: "We will make the landscape better than it ever was before." Fusing dated understandings of primitive culture and pristine nature, these industrially designed "Indigenous environments" are now celebrated by the oil industry as evidence of both the moral character of the corporation and a technical endpoint of the present degradation of the region. I take these corporate funded "Indigenous environments" as patently false and deeply tied up with the epistemological labor of purification and silencing that critical scholars and Native leaders have long rejected as colonial through and through (Smith 1999; Coulthard 2014; Simpson 2014; Estes 2019; Estes and Dunbar-Ortiz 2020). What interests me here is not so much the empirical falsehood of these "Indigenous environments" as their theological certainties. Describing the culturally sensitive reclamation projects he oversees, Steve continued: "First Nation communities have no idea how good they have it with us." For company officials like Steve, these projects to build a premodern future offer a compelling vantage point on present operations, one that simultaneously recognizes and redeems the disastrous qualities of tar sand extraction.

From afar, the critique of the tar sands often centers on the cinematic scale of devastation that marks the extraction process. "Hiroshima" was the phrase Neil Young lobbed in his explanation of all that is wrong with the tar sands. Others have struck a similar chord, describing the tar sands as "industrial Mordor" or "post-apocalyptic wasteland." After spending a portion of two summers in northern Alberta, to me such critiques seem to miss their mark a bit.[1] It's not so much that the portrayals are inaccurate—the scale of destruction *is* shocking—as that the companies have already conceded the point and are now presenting themselves as the best-equipped organization to rebuild the worlds they themselves are destroying. This chapter is an attempt to work out a slightly different critique of the tar sands. Rather than suggesting that exposing the disaster constitutes critique, I want to draw attention to the ways in which the current recognition of the disaster works to expand the moral authority of the corporation into new cultural and environmental realms—and, in the process, works to negate rising dissent from First Nations. This dynamic paradoxically celebrates an industry-endorsed theory of the "Indigenous environment" as a potent means of silencing the voices and demands of First Nations working tirelessly to protect their homelands and lifeways against the incursions of oil. A corporate theory of the "Indigenous environment" lies at the heart of this chapter: the extractive practices it authorizes alongside the ongoing colonial violence it both turns its back upon and charters anew.

"Moral economies," with a few revisions, offers a productive frame to unpack these concerns. Tracing the rise of capitalism in England, E. P. Thompson (1971) describes how rural communities drew upon local customs and beliefs to confront the social turbulence of profit and hold it in check. Whether in boisterous riots over bread prices or in subtle refusals to concede the commons, these "moral economies" offered a popular and potent counterpoint to the abstract logic of the market. They also signaled how a rooted understanding of right and wrong stood outside of the circuits of capital, enabling new points of friction, negotiation, and resistance. Such "weapons of the weak," as James Scott (1985) famously described them, became a robust rallying point for progressive scholarship and social movements looking for local alternatives to capitalism. Early formulations of moral economies, however, caught something later celebrations have neglected: in moral economies, traditional norms are not a timeless given but are selectively realized in confrontations with new regimes of exploitation. The confrontation, we might say, distinguishes past from present and sharpens what came before into an effective protest of the present.

Today, the presumption that moral economies might offer a critical limit to capitalism is no longer an obvious or easy point to make. The creativity of capitalism, as Luc Boltanski and Eve Chiapello (2007) have recently argued, lies precisely in its ability to transform moral counterpoints to capital into moral technologies of capital. Corporate social responsibility—the "new visible hand" of capitalism, as Dinah Rajak (2011: 11) puts it—is co-opting the very practices and locations of moral economies (and actively constructing more pliable versions of them where they cannot be found). This is especially true in extractive industries. Describing how matters of cultural heritage, local values, and even discourses of indigeneity itself are fast becoming internal to corporate mining strategies, Rajak (2011) writes: "Moral economies of responsibility, generosity, and community—and the social bonds of affection and coercion that these create—have become *not* the weapons of the weak but the weapons of the powerful." These insights have become potent ethnographic questions, as it is the ethical dimensions of extractive industries (many of them oil companies) that most effectively help withdraw the floor of citizenship, redistribute risk downward, and grease the skids of exported wealth (Ferguson 2005; Welker 2014; Dolan and Rajak 2016; Rogers 2015; Appel 2019). As Stuart Kirsch (2014) has noted, mining industry journals now regularly encourage extractive industries to incorporate "aboriginal communities" as stakeholders in order to defuse their potential as critics. In northern Alberta, many oil companies work to ensure the empirical content and critical capacity of the "Indigenous environment" no longer stand outside of infrastructures of energy extraction but are made integral to the very operations of those infrastructures.

As the agenda of moral economies has doubled to encourage compliance alongside resistance, the temporal frame has doubled as well. While E. P. Thompson and James Scott located moral economies in lingering memories of different ways of doing things, today in northern Alberta the "Indigenous environment" increasingly draws its moral force both

FIGURE 6. Holding ponds in tar sands of Alberta. Photo by author.

from the alternative past it recalls *and* from the alternative future it forecasts. Remediation plans at Shell Oil's Jackpine Mine, for example, are built around using ancient aboriginal understandings of the land to engineer "self-sustaining ecosystems that are comparable or better than what was there before our development," as the impact assessment plans outline. In interviews, some company officials even suggested that by removing the bituminous sands from the landscape, oil companies are actually purifying the boreal ecosystem so that it can finally align with timeless Indigenous conceptions of it. In these ways, the oil companies present themselves as doing a great service to First Nations: tar sands operators are extracting the lifeblood of modernity—fossil fuels—from the landscape so that it can become attuned to a premodern cosmology of it. These selective investments lead to a paradox commonly experienced in the tar sands: oil companies routinely invest tremendous sums to better understand and care for the "Indigenous environment" of the past and future while at the same time neglecting the rising chorus of Indigenous voices concerned about their environment in the present. The temporal outside provides a moral high ground for critics and proponents of the oil industry alike.

My point here is neither to undermine the real crisis of toxicity that assails First Nations nor to dismiss the critical potential of cultural differences, but to take note of the strange convergence of investments in an "Indigenous environment" that seems to gain moral authority from its conceptual distance to our present. Social theorist John Urry (2009) has recently called for a "post-carbon sociology" that would shift critical attention from the problems that assault our troubled present to the alternative futures we all must seize upon if we are to survive. As just such an alternative future, the "Indigenous environment" has become a key rallying point for advocacy groups looking to disrupt oil and corporations seeking to extract the oil in a more responsible manner. Yet the "Indigenous environment" as it is currently celebrated by activists and corporations often sidesteps the active presence of colonial history and the creative capacities at work in First Nation communities today.

FORMALIZING THE IMPACT

Tar sands extraction is not easy. The boreal forest has to be bulldozed and the entire landscape peeled off to expose the thick deposits of bituminous sands below the surface. Gargantuan bucketwheel excavators then swing back and forth along the exposed dark earth, eating their way ever forward and loading oily sands on giant conveyor belts or oversized trucks.[2] Removing the bituminous deposits from the sand is an elaborate operation that requires massive inputs of water and energy.[3] Tar sands operators are allocated roughly 8 percent of the total flow of the Athabasca River, and most of that water is used to wash the tar sands. Stadium-sized washing machines boil the bituminous deposits in water and petrochemicals, effectively separating the sand from the hydrocarbons. A refinery then mixes the separated bitumen with solvents to make them more amenable to pipeline travel while the wastewater—now laden with sand, heavy metals, and petrochemicals—is pumped into sprawling tailing ponds. By surface area and by volume of construction material, these ponds form some of the largest reservoirs on Earth. They are also toxic, and elaborate ruses like mannequins and propane guns warn birds away from their surfaces. These ponds are designed to fix the contamination in place; they are supposed to sit for centuries while toxic chemicals in the wastewater stabilize and settle in the sediment below the surface. But nothing stays put.

Almost from the moment the industrial extraction and processing of tar sands commenced in the 1960s, First Nation communities in the region felt the impact. Fort McKay, today home to about seven hundred tribal members, is perched above a bend in the Athabasca River adjacent to tar sands operations. In the 1960s, "the community began to notice that drinking water from the Athabasca River induced nausea and vomiting" (Longley 2015: 212). While the oil companies and the provincial government neglected to consult with Fort McKay about the encroaching operations, the Alberta Department of Health did warn the community to avoid drinking river water as tar sand operations got underway. As environmental historian Hereward Longley (2013: 113) writes, "By 1980, residents of Fort McKay reported that they could no longer wash clothes with river water because they would stick and cause skin irritation. The community reported that pike and pickerel caught from the Athabasca River tasted bad and induced vomiting." Fish kills became a regular occurrence, and rainwater came with a yellow scum.

Through the effective mobilization of First Nations regionwide and a number of lawsuits, by the 1980s Fort McKay compelled tar sands projects that infringed upon its traditional fishing and hunting areas to formally engage the village. Fort McKay, like many First Nation communities around the tar sands, gained title to their land in Treaty 8. Signed in 1899, Treaty 8 was informed by sharp Native discontent with previous treaties on the plains that confined First Nations to reservations (Fumoleau 2004: 79). Against those experiences, negotiations ensured that Treaty 8 provided both sizeable allotments of land to First Nation communities and the unrestricted right to trap beyond those formal tracts.[4] Those latter rights—the right to hunt and fish in the open forest and lakes—have become the crux of how oil companies have been compelled to negotiate with Native communities and incorporate the "Indigenous environment" into their plans.

"Fort McKay is very unique among First Nations in the area," an official in the tribal government told me in 2014. The Fort McKay First Nation has land reserves and traditional hunting rights atop some of the richest tar sands deposits in the region. As both cultural impacts and traditional ecological knowledge were added to the environmental impact assessment required to permit new projects (Goldman, Nadasdy, and Turner 2011), nearly every oil company operating in the tar sands found itself obligated to consult with the Fort McKay tribal government. This consulting quickly overwhelmed both the tribal government and the oil companies. As it was explained to me: "The Chief got sick of companies just showing up at his door at all hours asking for information on historical and traditional use sites." Instead, the tribe decided to write its own standardized account of its traditional uses of the land that could then be licensed to oil companies. At the time, many oil companies were equally overwhelmed by the requirement to consult every potentially impacted First Nation community. One company report from 1996 notes: "The sheer number of applications currently under-way in the Oil Sands Region of northeastern Alberta makes it almost impossible to update traditional uses of the land for every application." The companies, too, wanted a standard definition.[5]

Fort McKay conducted its own study of its traditional hunting and fishing practices beyond its formal land reserve and the relation of those hunting and gathering practices to the community. It decided that the date 1964 would be the benchmark for its definition of its traditional self. An early version noted: "The date was selected because this was prior to large-scale oil sands development within Fort McKay Traditional Lands" (cited in FMIRC 2010: 4). This was not a stable point by any means, but one that minimized other changes to focus more specifically on the threats that the arrival of large-scale oil extraction posed to their hunting and fishing activities. For example, in the 1950s the federal government began mandating residential schools and village housing as a policy to more forcefully push northern First Nations out of the forests and on the pathway to wage labor. Forts, previously company towns visited seasonally by many First Nation communities in the boreal forests, became a more permanent basis of tribal life. Many families no longer lived along their trap lines but instead visited them more periodically between the demands placed on them in settled life. It was these not quite recreational but not quite fully nomadic hunting and fishing trips that became the basis of the "tradition" most at risk by oil development. Fort McKay chose 1964 not to represent its timeless way of life but to pin down the specific practices, values, and aspirations that unchecked oil development had thrown into question. And yet as this understanding of Fort McKay began to circulate in company impact statements and remediation plans across the region, this knowledge lost its specific historical relation to oil. This baseline became not the reflexive response of one community to hydrocarbon encroachment but a key marker of the ahistorical authenticity of Indigenous life (Nadasdy 2005; Bessire 2014; Cepek 2016). Extracted from the contingent collisions and concerns that gave rise to it and editing out the longer (and ongoing) histories of colonialism, a selectively abridged version of Fort McKay's self-study of its subsistence activities in 1964 is found in nearly every environmental impact assessment on file in the Fort McMurray official tar sands repository. Spending days and days in the library, it took me a while to see this embroidered history of endorsed vulnerability in the region. It was only by reading older environmental impact assessments that I could see the editing process at work, as select subsistence activities in 1964 were extracted from history and flattened into a standardized definition of the timeless "Indigenous environment" for oil companies. Such work was not restricted to retrospective. This corporate theory of cultural ecology now provides a template for the engineered future, and it is from within the resulting eschatology that citizens, regulators, and stakeholders are encouraged to provincialize the present destruction.

This endorsed theory of the "Indigenous environment," however, has not kept toxic problems at bay. Now hemmed in on all sides by tar sands extraction, residents of Fort McKay are once more vocalizing concerns with air and water pollution and collapsing local animal populations. In 2006, an adjacent upgrading facility released a cloud of ammonia that floated over the village and sent a handful of kids to the hospital with breathing difficulties. The tribal government, attentive to the increased frequency of such events, created a new office of sustainability to help put in place more robust environmental monitoring. There has been tremendous resistance from the state and oil companies to these plans, often along the questionable lines that there are already robust environmental monitoring systems in place (during the 2006 ammonia gas event, company-run air monitors nearby experienced a software malfunction and were temporarily offline during the release).

DOWNWIND AND DOWNRIVER

Home to eight hundred people, Fort Chipewyan sits on a granite bluff above Lake Athabasca, looking out at the Peace-Athabasca Delta. Encompassing twelve hundred square miles, this convergence of rivers and wetlands forms the largest freshwater delta in North America. Its natural abundance is unrivaled: it is home to a dizzying array of wildlife, including the last wild buffalo herds in North America, and over a million migratory birds pass through each year (Timoney 2013). The area has long been occupied by Native people. Fort Chipewyan was established in 1788 and soon became one of the most lucrative fur trading outposts for its proximity to the abundance of the Peace-Athabasca Delta. Today, this remarkable ecosystem is shifting from a wellspring of subsistence to one of toxic suspicion.

About one hundred miles downriver of the oil fields, Fort Chipewyan has experienced much of the environmental fallout of the tar sands while reaping few of the economic benefits. Although many of the residents work for oil companies in some capacity and a few have even started their own companies, the community itself has been marginalized. By and large, what benefits have accrued have been to individual households, not at the tribal level. Since most of its traditional hunting grounds are well to the north of the oil fields, companies are not always obligated to consult with the tribal government of Fort Chipewyan. Yet runoff from their operations routinely impinges upon the hunting and fishing areas of residents. "There was a big fish kill a few weeks ago that kind of spooked us," one resident told me. Without investigating, state officials said it was probably just lack of oxygen unrelated to mining operations. But residents suspected oil.

Over the past decade, the Lake Athabasca fishery drifted into uncertainty as evidence grew that fish populations verged on collapse. While no specific cause has been identified, even state officials now warily acknowledged that runoff from the oil fields upriver probably played some role. Even before the fishery was closed, most people I spoke with in the village had stopped eating local fish. "I don't eat the fish, I don't drink the water," one resident said, echoing popular new fears of the rivers and lakes. Local mammal populations are in steep decline as well. And what can be caught locally now elicits more anxiety than celebration. The day I arrived the town was in mourning. One more tribal member had just died of cancer, and the whole community came out for the funeral procession. "He was an active guy, fishing, and trapping his whole life. He was one of the last still living off the land," I was told in a bar that night.

In 2003, a community doctor, John O'Connor, grew concerned with rising autoimmune diseases in the community and new cases of rather obscure cancers. After a few calls to state agencies went unanswered, Dr. O'Connor shared his concerns with the media. He was rather quickly rebuked by government and company officials who claimed his "political agenda" had distorted his medical obligations. In a 2006 presentation to the community, the Alberta Health Department noted the cancer rate in Fort Chipewyan was about 30 percent higher than the average for the rest of Alberta (cited in Timoney 2007: 6). Nonetheless, the state declined to conclude there was a cancer cluster in Fort Chipewyan, calling for further study instead.

The community and regional health board contracted their own scientist to offer a preliminary analysis of the Athabasca River and Lake

Athabasca. That study found high levels of toxins associated with tar sands mining in the rivers and lakes that surround Fort Chipewyan. Identifying polycyclic aromatic hydrocarbons (PAHs), arsenic, and mercury in these waterways, the study concluded "concentrations of these contaminants, already high, appear to be rising" (Timoney 2007: 4). The study also found rising rates of deformities in fish associated with declining water quality, and it raised concerns about potential pathways for dangerous human exposures through subsistence activities in Fort Chipewyan. The author recommended the community organize a more formal research program into these concerning findings: "An environmental monitoring program independent of control by vested interests is needed. The program should be affiliated with a university and report regularly in open public forum to the people of Fort Chipewyan" (Timoney 2007: 5). Two comprehensive studies released by the *Proceedings of the National Academy of Sciences* confirmed these growing suspicions: mining operations in the tar sands were dumping a tremendous amount of known carcinogens into the waterways that many First Nation communities depended upon (Kelly et al. 2009; Kelly et al. 2010). Petrochemicals from mining operations were detected regionwide in insects, sediment, and even snow (Wayland et al. 2008; Kelly et al. 2009; Timoney and Lee 2009).

Industry representatives and provincial officials continue to downplay questions of the petrochemical fallout of mining operations and evidence of a cancer cluster in Fort Chipewyan. In 2009, a health survey of Fort Chipewyan conducted by the Alberta Cancer Board concluded, "The number of cancer cases overall was higher than expected" (Chen 2009: 10). Documenting fifty-one cancer diagnoses in the village of seven hundred, where only thirty-nine would be expected, the findings warrant "further assessment and closer monitoring" of Fort Chipewyan, the Alberta Cancer Board concluded, including further attention to "environmental exposures" (Chen 2009: 10). A few months later the state disbanded the Alberta Cancer Board. That same year, the state quietly pushed to have Dr. O'Connor's medical license revoked. The College of Physicians and Surgeons instead rebuked Dr. O'Connor for making "inaccurate" claims that caused public panic and "obstructed" scientific monitoring of the community's health (CPSA 2009: 2). While university scientists continued to map out just how vast and injurious the contamination was, the provincial government continued to assert that contamination from industry was not detectable in the Athabasca watershed (Alberta Government 2010). Several scientists noted the existing government program to monitor water quality on the Athabasca River was focused entirely on contaminants unrelated to tar sands mining and was "unable to measure and assess development related change" (Timoney and Lee 2009: 2).[6]

In early 2014, a health study supported by the oil industry and the provincial government concluded there was no elevated cancer risk. James Talbot, the chief medical health officer of the Alberta Health Service, summarized: "Overall the cancer rates are what would be expected for the rest of Alberta" (Weber 2014: 1). Chief Allan Adam of Fort Chipewyan retorted that the community was still waiting on "a comprehensive, independent study," one that "industry have no participation in it" (Weber 2014: 1). That same year, additional research began to document a deluge of contamination leaking from tailings ponds across the region. A single tailings pond was found to contribute nearly two million gallons of waste water laden with carcinogens into Athabasca River each day (CBC News 2014). It is located just upriver of the Peace-Athabasca Delta and Fort Chipewyan

Taking matters into its own hands, the Mikisew Cree First Nation and Athabasca Chipewyan First Nation partnered with university-based scientists outside of Alberta to conduct their own investigation.[7] In late 2014—after a $1 million, three-year study—this research project released its findings, documenting alarming pathways of exposure that linked tar sands pollution to cancers in Fort Chipewyan (McLachlan 2014). Analyzing levels of petrochemicals in fish and game caught in the Peace-Athabasca Delta, the study found high concentrations of carcinogenic PAHs and toxic heavy metals like cadmium, selenium, and mercury in the meat of locally harvested moose, ducks, muskrats, and beavers. Surveying the community for the cancers known to be caused by exposure to these toxins, the study identified twenty-three cancer cases in Fort Chipewyan likely related to oil pollution: "For the first time, we showed that upstream development and environmental decline are affecting cancer occurrence. Thus, cancer occurrence increased significantly with participant employment in the Oil Sands and with the increased consumption of traditional foods and locally caught fish" (McLachlan 2014: 12). In interviews, the authors of the report issued warnings about consuming locally caught fish and game. The lead author, Dr. Stéphane McLachlan, noted, "People who consumed traditional foods more frequently and those who consumed locally caught foods were more likely to have cancer" (Edwards 2014: E444).

The manager of the single grocery store in Fort Chipewyan told me he has a hard time keeping his shelves stocked with bottled water as demand skyrockets. Fresh vegetables were even harder to come by. Rising fears over contaminated animals, fish, and water have transformed the waterways and boreal forest that surround Fort Chipewyan from a source of life into its opposite. The land itself has become haunted by the disasters of hydrocarbon extraction. These toxic concerns take what was once the explicit policy of colonialism—forcing Native people off the land and into wage labor—and imbue it with new medical justification, amplifying the violence of colonialism almost independent of the heavy hand of the imperial state. These concerns have also led to a new push for the scientific equipment and expertise to help the community better understand what might be moving through their water and air and animals, as well as new efforts to mobilize around what has been lost. For many residents of Fort McKay and Fort Chipewyan, the negative ecologies of the oil industry exceed any ledger of gain, derail any promise of progress, and conspire against any sustainable hope for health.

After I had a few email exchanges with the director of aboriginal relations at a major oil company in the tar sands, he invited me out to dinner to talk about "the Fort Chipewyan problem." We met at an upscale restaurant in Fort McMurray, where he introduced himself to me as a "total foodie." He and his wife "vacation the Michelin List," he told me, each year visiting a handful of the most prestigious restaurants on earth. The previous year he had secured coveted seats for dinner at El Bulli, and the pilgrimage had changed him. Laughing, he explained how he now had "laboratory grade emulsifiers" crowding his kitchen among other scientific instruments repurposed for high culinary ends. A dinner invitation, he hinted, might be in the works should we reach an understanding. He also let it be known that his staff had looked me up and read my publications and had debated that very afternoon about whether I was political. He was, he said, happy to talk to scientists but had no time for people with a political agenda.

I explained my interests, saying I wanted to understand how oil companies brought environmental science and traditional knowledge together to better manage the impacts of their extractive operations. I said many environmentalists and First Nation communities suspected that remediation projects were just good advertising, a thin window dressing to cover over what was really going on. But it seemed like oil companies were investing substantial sums of resources and personnel into these projects, and I wanted to understand why. "What do oil sand companies hope to get out of these investments?" I ask. He explained the genuine interest oil companies now had in working with First Nations. "I'll be the first to admit, we made many mistakes in the past." First Nations had neither been consulted about nor compensated for the real disruptions of tar sands extraction. But oil companies had learned their lesson. Negotiating with First Nations was not only the right thing to do, it was also good business. He struck me as utterly sincere on this point: learning to listen to First Nations had improved the oil industry. It had provided real opportunities to collaborate on improving the design, operation, and remediation of tar sands mines and tailing ponds. Today, Native-owned corporations were prized partners in tar sands operations, and the ecological wisdom of First Nations oriented the goals of environmental restoration.[8]

As we ordered our meals, I shifted the conversation toward the scientific reports of significant downwind and downstream pollution from the tar sands and its potential impact on human health. He rebuffed the implication: "There is no evidence that its coming from our operations," he said, before describing the miles of riverbanks where bituminous sands naturally seeped into the Athabasca River. Mining those sands, he noted, might be cleaning the river up.[9] As the waiter set down our plates, I asked him about the cancer reports. With a sigh, he explained how much his company had done for Fort Chipewyan. "We paved all their roads for them. You don't see that much in isolated communities, do you? And did you see the new elder center they're putting in? We're helping out with that." And yet, he noted, the community remains ungrateful. "The chief has a relentless political agenda."

I asked about the collapse of animal populations in the region and the relative dearth of fresh food in the isolated village. "They wouldn't actually eat fresh food. They actually like the packaged stuff better," he said. I asked about cancer again. "Well, what causes the cancer up there? I'll tell you what causes cancer in Fort Chipewyan. One, sedentary lifestyles. Two, smoking. Three, bad diet. Four, sexual promiscuity. It's all behavioral," he said as the waiter brought desserts decorated with candied spirals and pansy flowers. "Oh it's exquisite, " he proclaimed, "this is as good as anything you'll find in Chicago or Houston."

MONITORING THE IMPACT

As frustration over toxic clouds and cancer clusters grew, the provincial government of Alberta decided to dismantle its current system of monitoring air and water quality in the region to build what it advertises as a "world class environmental monitoring program" (JOSM 2016). In the short term, this has meant that ongoing monitoring was paused in anticipation of the new and improved system. The lone state employee tasked with environmental monitoring in Fort Chipewyan has been asked to stand down. "I'm gainfully employed," he told me in 2014, "but I'm doing nothing, I have no work duties for the indefinite future." Bored out of his mind, he wrote the main office in Calgary to ask if he might keep up the water sampling and tests he had already started on his own time and at his own expense. He showed me their emailed response, which stated quite bluntly that he would be terminated if he collected any environmental data. And he was no critic of the tar sands. "I'm not political," he said, and "we don't yet have the data that proves the tar sands are causing any harm." In fact, he noted, all the critical attention that had condensed around the tar sands had turned a blind eye to the other toxic problems seeping into the Athabasca watershed: namely, massive paper pulp plants and abandoned uranium mines. Nobody's even looking for what they might be putting in the river, he said.

In 2013, Alberta decided to consolidate all of the federal and state environmental oversight of tar sands projects into a single agency. Moreover, this agency would be run at no expense to taxpayers, as the Canadian Association of Petroleum Producers agreed to foot the bill. By fully funding this new and improved agency, however, the oil companies played a lead role in designing the new rules and regulations for that agency. The problems of this arrangement became apparent in the negotiations over the specifics of environmental governance.

The new regional system for monitoring air and water quality and animal populations in the region is called Joint Oil Sands Monitoring (JOSM). This system promised a comprehensive environmental monitoring not of individual projects but of the entire region. The consequential details of how this monitoring would take shape were to be ironed out in negotiations between the provincial and federal governments, oil companies, and First Nations leaders over a three-year period ending in 2015. The mission was clear: the program was to establish data points to measure air and water quality, animal populations, and other ecological indicators not at the level of individual extractive projects but across the entire region. From that geography of measurement, the program could begin to see "the relative health of the area's ecosystems" and formulate the "baseline yardsticks against which all future data will be compared." Although it is a bit peculiar to establish the comprehensive environmental baseline for the region in the midst of rampant expansion of mining operations, other problems emerged as well.

From the get-go, First Nation leaders pushed to be involved in the technical details of what specific pollutants to measure, where to measure them at, and how to set thresholds for toxicity. "We were very clear in what we wanted," one tribal leader from Fort Chipewyan noted. "We wanted to be involved in the decision-making and design" of environmental governance (McDermott 2014a: 1). More specifically, they wanted to make sure that air and water sampling took place around their communities. Again and again, these demands were rebuffed by government and company officials. First Nation leaders, one local paper reported, "felt they were being pigeonholed into providing solely traditional and cultural knowledge" (McDermott 2014a: 1). Before the three-year process of determining the methodology and distribution of environmental monitoring in the tar sands was complete, all five First Nation groups withdrew from the program in protest, including Fort McKay and Fort Chipewyan. Each complained of the strange resistance the program had to instituting regular air and water quality monitoring in First Nation communities (McDermott 2014b).

Companies, it might be said, appeared to be more interested in safeguarding Indigenous traditions than Indigenous lives. While making tremendous investments to understand and safeguard the now timeless "Indigenous environment," the same companies seem actively uninterested in First Nation communities' environmental concerns today. Many impact assessments and reports on corporate social responsibility in the tar sands contain some variation of this line found in a recent application for a new mine: "Aboriginal ecological knowledge is the wisdom and understanding of a particular natural environment that has accumulated over countless generations and can serve to aid Western scientific disciplines in analyzing project impacts." By providing an cultural ecological baseline to measure the impacts of the project against, traditional knowledge of the land is used to define the temporal and ethical boundaries of the disruption of tar sands operations. As companies find ways of dodging and disabling contemporary concern over environmental protections in First Nations communities, those same companies paradoxically celebrate the "Indigenous environment" as the starting point and ending point of their obligations.

The "Indigenous environment" works as a novel moral economy within the oil industry today, one that allows for the disasters underway in the present by insisting on apprehending them from their future redemption. If the negative ecologies of the oil industry cannot be infrastructurally contained, perhaps ontology can better anchor the destruction in place. I take these "Indigenous environments" not just as a convenient cover for baser interests but also as a consequential new mechanism for sorting out the legitimate vulnerabilities and the limits of moral obligations generated by the destruction of extraction. While those practices and subjects that align with the narrow corporate interest in the "Indigenous environment" are showered with exceptional care, those falling outside that narrowing vision are treated with new forms of acceptable disregard (Bessire and Bond 2014). Corporate investment in the "Indigenous environment" is a way for companies not only to acknowledge the risks of their tar sand operations but also to delimit their responsibility to them.

THE REAL COST OF ENERGY

"The real cost of energy": it is an unobtrusive thought, a phrase noted in passing throughout the tar sands. It's affixed on the banal surfaces of the company, almost like common sense, almost like a warning. It's sounded out in press releases and interviews. And it is only ever half registered. "This is the cost of energy today." One sign hanging on the wall of the welcome center in Fort McMurray, Alberta, reads: "When we use energy from the oil sands to make our lives more pleasant, it has an impact and an environmental cost. [...] We cannot use energy and have no impact at all." It came up routinely in interviews with company officials. As I steered our conversation toward the environmental dimensions of surface mining, they would gently ask: "Do you drive a car?" or "Do you heat your home in the winter?" or "Did you fly here?" As if such questions answered themselves. As if the vast devastation of the landscape is entirely unavoidable for those who expect the creature comforts of modernity. Not only did this slogan help diminish the ecological impact of the tar sands into subservient relation to broad technological uplift, it also worked to distribute responsibility for that devastation equally to everyone who uses an automobile or desires a warm home in winter. If we are all liable, no one can be prosecuted. In the Oil Sands Museum, one exhibit features an aerial photo of a sprawling tar sands mining operation. The inscription beneath the wall-sized image asks: "Who is responsible for protecting the environment? (A) The Government; (B) Oil Sands Companies; (C) Everyone." The correct answer, the fine print explains, is everyone.

A few tar sands operators provide guided tours of their mining operations. Before the tour starts, a scripted narration anticipates the destruction that they will be witnessing. "When we use energy from the oil sands to make our lives more pleasant, it has an impact and an environmental cost. New research can help us avoid some of the problems and limit others, but we cannot use energy and have no impact at all." Although it begins at the mine, the tour ends at a restoration site where, as the guide explained, Indigenous wisdom, engineering wizardry, and the abiding commitment of the company join together to transform a tailings pond into a more perfect native ecosystem. "Renewing the landscape" was the phrase the guide used. We were told how plants were carefully selected by tribal elders, then propagated and planted en masse by environmental scientists. And we were told how the traditional knowledge of First Nations helped map out places for piles of rocks and dead trees that would help bring back the biological and spiritual life force of the landscape.[10]

Suncor, a major tar sand operator, recently finished work on the "Crane Lake Reclamation Area" as a technical prototype and shining example of its commitment to restoring the "Indigenous environment." The timeless traditions of the First Nations played a key role in the design of the area, he explained. This site, a director of remediation at Suncor told me, was designed in consultation with Indigenous elders in the area to "make it better than it was before." While gleaming photos of "Crane Lake Reclamation Area" abound in the company advertisements and government reports, few people ever seem to visit the place. It is a strange place, sandwiched in between two sprawling refining complexes. Smokestacks loom over the engineered forest, while the air is filled with the constant thud of explosions to scare birds away. A company employee standing guard nearby explained the elaborate system of water cannons, lasers, and scarecrows that work to dissuade birds from landing in ruins of the tailing ponds. Landing on the pond, he admitted, usually killed them. The whole place smelled of hot asphalt. At the end of a short trail, a wetlands area has been designed with input from environmental scientists and Indigenous elders to, a sign explained, "attract native animals and birds back to the area," so that this landscape might once again support subsistence activities. A sign near the entrance reads: "Please do not disturb the Eco System. Absolutely no diving, no swimming, no wading, no fishing is permitted for your safety."

Syncrude manages a small buffalo herd atop one reclaimed tailing pond. A parking lot and an engineered hill provide a viewpoint to watch the modest herd of wood bison grazing in the pastures between an active mine and a mammoth pile of sulfur. (I asked a guide if anyone harvests the buffalo. No one would eat those buffalo, a local resident chimed in, "Look at where they are.") Billboards along the viewpoint ask visitors to imagine the buffalo one day running wild on this redeemed landscape. "The Future—A Whole New Point of View" reads one sign. From the selective lens of a brochure or corporate common sense or even regulatory oversight, these sites can appear as a perfect picture of the Indigenous environment. Up close, however, we can see that these selective and strategic images of the future rest atop a toxic stew of still-simmering calamity.

The work being done is more than advertising. These reclamation sites—and each major tar sand operator seems to have at least one—are designed to imagine the landscape post-oil. Many are built with asphalt walking paths and parking lots capable of hosting tour buses. The resulting vision of ecological alterity does crucial temporal work for the oil companies. These company-sponsored "Indigenous environments" are used by oil companies to condense the wound they are inflicting on the landscape into the fixed dimensions of a "modern event." That is, the "Indigenous environment" provides the essential before *and* after. The present destruction is displaced by the promise of what is to come. Active mines visible from the highway have billboards along them: "Reclamation in Process." Present Native communities are denied a meaningful role in environmental governance while, at the same time, the "Indigenous environment" is hoisted up as the premodern past and premodern future that mark out tar sands destruction as a bounded, fixed, and entirely passing event. While destruction becomes accountable as a secondary cost, the negative ecologies of the oil industry are rendered illegible in the outlook of this modern event.

In the "Environment" section of the Oil Sands Museum, a four-minute video plays on a loop. Panning shots of the boreal forest at twilight open the video as a soft flute plays in the background alongside the slow strumming of an acoustic guitar. The video, produced by a consortium of oil companies, describes how energy makes our lives possible, and energy has a cost. A female narrator explains that today oil companies have expanded their operations to better manage and minimize that cost. "Oil sands companies are paying a lot more attention to aboriginal elders and their understanding of how animals, plants, and people interact in the boreal forest," she says.

Fred McDonald, a First Nation leader interviewed for the film, talks of declining water quality and ecosystems coming undone as tar sands mining expanded in the past decades. "Everything around here has pretty well been disturbed." McDonald describes the new absence of abundance. Sightings of moose and deer have become exceptional events, geese and ducks pass through without stopping, and "robins don't stick around anymore, the blackbirds don't stick around anymore." McDonald is clear, the arrival of oil sands extraction set these disruptions in motion. The oil operations, McDonald says, "chased everything out of the area." "Information from elders," the narrators continues, "can give environmental scientists insight into what is changing in the natural history of the Fort McMurray area as well as hinting at some of the reasons for those changes."

Describing wave after wave of impacts, McDonald ends his litany with water. "Water is getting to be more important than oil. If we destroy the water, we're all in trouble, not just the animals." Throughout the video, McDonald speaks of loss without remedy. "Things is going to get worse, very bad, instead of better." As if to finish the thought, the narrator steps in: "As Fred notes, we can't produce oil from the oil sands with no costs or consequences at all." But with the right science and engineering, the narrator continues, the oil companies can rebuild "the kind of balance and good stewardship of the environment that Fred McDonald's traditions honor."

In 2018, a high-ranking official in the Alberta Energy Regulator (AER) dropped a bombshell: the cost of cleaning up the tar sands may actually exceed the market value of the tar sands. Rob Wadsworth, vice president of closure and liability for the AER, is tasked with calculating how much tar sand companies should be setting aside to ensure full restoration of the region. Reviewing the accounting for an internal report, Wadsworth realized that the existing rules grossly underestimated the actual costs of full restoration, provided a century-long horizon of inaction, and allowed for abandoning wells and drilling sites once the oil ran out if the companies could no longer pay (DeSouza et al. 2018: 1).[11] Indeed, in 2016 AER noted an uptick in wells finding their way into the "Orphan Well Fund," meaning tar sands operators had abandoned them without fully remediating the site (Johnson 2016). If those detractions and deferrals were zeroed out, Wadsworth wondered, what would the actual cost of restoration be? Wadsworth found outstanding liabilities for full restoration to be on the order of $260 billion. As environmental

FIGURE 7. Billboard announcing the remediation to come: "Reclamation in Process." Photo by author.

historian Hereward Longley (2019b: 1) points out, that ecological debt grossly exceeds the combined market capitalization of the five biggest tar sand corporations, which clocked in at $165 billion in 2019.[12]

CONCLUSION

While fossil fuels are activated within relations of capital (and in turn accelerate the relationality of capital), the consequence of fossil fuels is not contained within those relations. Whether as explosive disaster or as slow violence or as planned-for contingency, the negative ecologies of fossil fuels routinely exceed the critical and normative analytics of the commodity form. Over the past century and already spilling into our future, fossil fuels have instigated new landscapes of vulnerability, new mappings of the mediums of harm, and new responsibilities to manage those domains. Such destruction has enacted the oppositional moral economies of sustainability and environmentalism that fossil fuel industries now seek to incorporate and extend as vital ethical technologies of oil itself. While the previous chapter described how an unprecedented deepwater blowout extended the authority of the state into new oceanic depths, this chapter has shown how the ordinary disruptions of tar sands extraction extended the authority of the corporation into new realms of cultural ecology. In both situations, the dissonant experiences of those living closest to these disasters is rendered nearly illegible by the powerful recognition of them. Ethnographic attention to this dissonance, and the moral and scientific fields that cohere around the disastrous materiality of fossil fuels, might offer a new lens to understand and confront the role of fossil fuels in the contemporary. Not only could this bring the gradients of human difference newly expressed in the horizons of petro-risk into clearer focus, it might also help us see the negative ecologies of fossil fuels not as accidents condensed in time and space but as the fertile soil within which new political theologies of energy are taking root today. The subsequent chapters explore what happens when recognition of these negative ecologies provides a counterpoint to the capacity of the state and the corporation.

CHAPTER 4

Occupying the Implication

Politics makes visible that which had no reason to be seen.

—Jacques Rancière (2001)

This chapter switches the direction of negative ecologies. While the first half of this book focused on how the destruction of fossil fuels became a new terrain for state authority and corporate responsibility, in the second half of the book this excess becomes something that trips up the conceit of power and profit. The possibilities of negative ecologies for new alignments of politics first became visible to me through participating in pipeline protests.

In the contestation over its approval, the Keystone XL Pipeline offers a telling window into the contemporary politics of fossil fuels in North America. Although oil pipelines have been around for a century, they have long been neglected in anthropology and public debate. Today, that is beginning to change. Whether as a strategic vehicle for energy independence or as an urgent front line in the fight against climate change, oil pipelines are increasingly understood not as inert things but as consequential projects in our troubled present. This chapter uses the promises and protests surrounding the Keystone XL as a prompt to reflect more broadly on questions on how negative ecologies exceed vital infrastructure, and how that surplus is becoming a new terrain of social change.

In 2015, I rented a car and drove the presumed path of the Keystone XL through Nebraska, South Dakota, and Montana. Owned by the TransCanada Corporation, the Keystone Pipeline System consists of four phases, the first three of which are already built and in operation. Phase one repurposed existing natural gas pipelines and built new pipelines to connect the Alberta oil fields to Winnipeg some seven hundred miles to the east before heading due south to a pipeline junction in Steele City, Nebraska. Phases two and three installed a brand new thirty-six-inch-diameter steel pipeline that linked the refineries of Houston, Texas, to the Steele City junction. The controversial Keystone XL seeks to extend this new pipeline from Steele City directly to oil terminals in Alberta, forming a sort of hypotenuse on the existing Keystone Pipeline System. As designed, the Keystone XL is really just a shortcut.

It is worth noting, the Keystone XL is far from the first or even the only conduit bringing tar sands oil into the United States. Not only have trains carrying crude from the tar sands become commonplace in many parts of the country, but a handful of major pipelines now carry bitumen diluted with chemical solvents, or *dilbit*, from Alberta to US refineries. In many cases, pipeline companies retrofitted or simply reversed the flow of existing pipelines to avoid the public scrutiny of a new project. The Keystone XL is unique in that, as a new border-crossing pipeline, it requires the US State Department to review its impact and attest to the pipeline being in the national interest.

The thousand-mile route of the Keystone XL cuts across the northern Great Plains, from windswept cattle ranches in Montana to lush farms in Nebraska. After crossing the Canadian border, the Keystone XL heads down to Baker, Montana, where its Canadian cargo will be joined by the current glut of domestic crude coming out of the Bakken Formation in North Dakota and Montana. Mindful of potential points of resistance, the pipeline then squiggles its way eastward through Montana and South Dakota in studious avoidance of Native American reservations. In Nebraska, the first draft of the Keystone XL drew a rather brash line across the Sand Hills and Ogallala aquifer. This route elicited such indignation from local farmers that TransCanada quickly adjusted it, now swinging the pipeline out along the edge of the aquifer. These changes to the route have been mired in the Nebraska courts ever since, as the law is unclear about who actually has the authority to change the plan.

Although Keystone XL steers clear of major cities, it does pass by about fifteen small towns. In contrast to what is so often reported from afar, in towns like McCool Junction, Nebraska (pop. 413), or Midland, South Dakota (pop. 127), or Circle, Montana (pop. 617), I met folks who are neither adamantly for nor against the pipeline. Beyond the green-clad pipelines being stockpiled in an empty field just outside Buffalo, South Dakota (pop. 380), the Keystone XL had little visible presence in these towns. There were no rosy company billboards, no yard signs expressing an opinion, and the sizeable social investments TransCanada is making locally did not exactly advertise their funding source. (TransCanada advertisements, however, clad the company in the folksy authenticity of these small towns.) When asked directly, many residents around the oil fields of Montana voiced their support for the project. while many residents in the Sand Hills of Nebraska voiced their discontent with the project. But overall, most people I talked with along the route were ambivalent about Keystone XL, keen to have some decent local jobs but wary of the gloss of big corporations, especially foreign ones.

Walking down neighborhood streets strewn with rusty equipment and empty lots, it was not hard to imagine where this hesitance might come from. Many of these towns first sprang up alongside the movements of people and goods, trying to capitalize on whatever happened to be passing by. These towns have stitched their history together with the debris of military pacification campaigns, immigrant settler trails, transcontinental railways, and most recently the interstate. They are frontier towns, with all the colonial overtones and social histories of catch as catch can such description implies. They've seen boom and bust before, both in the feverish future such adjacent traffic promises and in the ruins so often left behind. Today, huge granaries fall into disuse along railways now crowded with coal trains not making any local stops. What commerce remains has drifted from a boarded-up main street to the single gas station on the outskirts nearer to the highway. In these ways and others, the towns along the Keystone XL route already bear the imprint of shifting material histories of transportation. They are not so much out of the way—indeed, what success they have had is owed to being very much in the way—as they are ever more meticulously passed by.

If the Keystone XL project is realized, a near astronomic amount of wealth will soon flow by these towns. And yet to an almost unrivaled degree, the wealth flowing through the thirty-six-inch-diameter steel pipe will accrue in concentration elsewhere (the risks, of course, will be widely distributed). Our vital economic networks don't seem to leak nearly as much as they used to, at least not of the stuff that might enable a community. Most of the relations of obligation engendered by this pipeline will be resolved in a single transaction.

About two thousand or so property owners will receive a check for a permanent right-of-way from the Keystone XL. Most will do so under the express threat of a novel and Supreme Court–sanctioned version of "eminent domain" that holds up corporate interest as a legitimate measure of the public good. Municipalities might see an uptick in tax revenues to fund long-needed school improvements and perhaps less-needed property tax breaks for residents. At least that's what communities have been promised if they don't live in Kansas, which in a legislative arms race to see who could appear more pro-pipeline—a one-upmanship that appears to have taken even TransCanada by surprise—has exempted all new oil pipelines from paying taxes at about the same time the state budget was slipping into serious financial turmoil. While local communities might see a flurry of workers desiring housing and food and lord knows what else during the construction boom, once built, the pipeline will be operated out of an office in Calgary. Keystone will require, at most, a handful of permanent workers in the United States. It is not until after the pipeline is buried and out of sight that the real wealth will start to flow, to the tune of about 800,000 barrels of crude oil a day. And of that most lucrative flow, local communities will see very little, if anything at all.

The Keystone XL, then, is an apt example of what political theorist Timothy Mitchell (2011) calls "Carbon Democracy." According to Mitchell, fossil fuels have played crucial if overlooked roles in shaping the practice of democracy today, both in its social aspirations and in its technical limits. At the dawn of the twentieth century, coal was at once essential to industry and quite laborious in its extraction and distribution. The dawning awareness of this reality joined disparate workers and empowered them to seize control of key chokepoints of coal in order to make broad social demands, an insurgency that compelled the companies and governments to expand their civic responsibilities. The extraction and distribution of crude oil in the post–World War II period moved in a decidedly different direction. Through imperial interventions, unmanned infrastructure, and oceanic shipping, the networks of crude oil worked to override earlier points of labored friction. In their design and operation, petro-systems have become quite adept at dodging and disabling robust forms of public accountability. For Mitchell, oil pipelines are the premier example of this progressive occlusion of workers and citizens. Driving through the small towns along the route of the Keystone XL in 2014, I could not help but reflect on how unremarkable this foreclosure of political possibility has become.

Two months later, in September 2014, I attended the Peoples Climate March in New York City. On the streets of that spirited and truly immense gathering, the Keystone XL had not foreclosed political possibilities but caused them to proliferate. Not only was an anti-Keystone XL message proclaimed in posters and rallying calls, but other pipelines like the Alberta Clipper in Wisconsin, the Exxon Pegasus in Arkansas, and even the proposed Cove Point liquid natural gas terminal in Maryland were being protested. One group of marchers wore baseball jerseys identifying themselves as "Pipeline Fighters." Another group, in a lively performance, dressed as a black hydrocarbon octopus whose sprawling tentacles became pipelines that chased wildlife down Sixth Avenue. While the Peoples Climate March lined up all variety of suspects for cathartic castigation, I was still taken aback by how ubiquitous pipelines have become in climate activism. A decade ago, I don't think any environmental activists in the United States could have predicted that pipelines would become such a rallying point.

Could climate activism be cultivating a new kind of social change? And might the growing presence of pipelines have something to do with this shift? By and large, the environmental movement in the United States has been a proscriptive movement, not a preventative one. Environmentalism gained moral and regulatory force in the outraged response to specific industrial disasters. Whether it was the suffocating smog of Donora, the flammable Cuyahoga River in Cleveland, the declaration that Lake Erie was dead, the smothered California coast during the Santa Barbara oil spill, or the domestic discovery of toxic waste in Love Canal, again and again it was spectacular events that catalyzed nascent environmental concerns and instigated change in policy to prevent their reoccurrence. In contrast, climate change demands public action not in reaction to an acute disaster but in anticipation of the diffuse disaster to come: a slow unraveling of the planet's climate. Although residents of the Maldives, Philippines, and coastal New Jersey might disagree, for many of us climate change does not yet have the density of a deeply felt event. As the novelist Zadie Smith (2014) points out in her essay, "Elegy for a Country's Seasons," climate change is a disaster that we have a "scientific and ideological language for" but "hardly any intimate words." How do we mobilize around the turgid prose of, say, the Fifth Assessment Report of the UN's Intergovernmental Panel on Climate Change? How do we effectively protest the predicted event? Where do we even begin?

A few years ago, a starting point in the fight against climate change suddenly emerged: the material infrastructure of fossil fuels. Although neatly sidestepping the more consequential question of what to do with those distinctly American lifestyles built on the presumption of hydrocarbon abundance, infrastructure nonetheless offered a practical place to begin. In 2011, the pent-up frustration of those incensed by the

FIGURE 8. 2014 Peoples Climate March in New York City. Photo by author.

lack of action on climate change rather potently aligned with the Keystone XL permitting process. It was NASA scientist James Hansen who first pointed out this convergence. In an open letter entitled "Silence Is Deadly" and posted online, Hansen (2011) noted that the window for public comment on the environmental impact assessment of the Keystone XL was about to close. He encouraged all concerned citizens to spread the word and log their discontent. "If this project gains approval, it will become exceedingly difficult to control the tar sands monster," he wrote. Extracting and burning all the hydrocarbons in the tar sands, Hansen argued, would push global warming well past the point of no return. If the Keystone XL was built, Hansen concluded, "it is essentially game over."

Bill McKibben, among many others, took note of Hansen's letter and got to work. McKibben and the organization he helped found, 350.org, have labored tirelessly to publicize the otherwise mundane process of permitting a new border-crossing pipeline in the United States. When that regulatory process didn't seem particularly newsworthy, they found creative ways of making it newsworthy, whether by writing vehement op-eds, overwhelming the public input process, or getting arrested at the

White House on prime-time television. Other national environmental organizations have followed suit, and today the Keystone XL is regularly described as the "lynchpin" to the tar sands or even as the "fuse" that will ignite climate change.

In Nebraska, a parallel resistance to the Keystone XL has taken shape, albeit one less concerned with how fossil fuels contribute to climate change than with how the Keystone XL might undermine regional agriculture. The organization Bold Nebraska has become particularly skilled at reframing the Keystone XL as an intrusion on property rights and thus a litmus test for state politicians: Do you support the pipeline or local farmers? Another organization, the Cowboy and Indian Alliance, has capitalized on the iconic convergence of interests that opposition to the Keystone brought about. At one Cowboy and Indian Alliance meeting, as Tony Horowitz (2014: 1558) reports in his book *Boom*, a rancher lashed out at TransCanada's strategy of eminent domain: "They're taking our land!" Realizing the disquieting history of the sentiment just uttered, the rancher added, "I guess that's what happened to you. Now it's happening to us."

Alongside this renewed activism, many online news platforms like ProPublica and Inside Climate News directed their talented investigative reporting teams toward the Keystone XL. National news outlets often found themselves following their lead. This burst of attention has revealed many embarrassing lapses in planning and regulatory documents that never anticipated such scrutiny. In an era of shrinking journalistic resources, this in-depth coverage of proposed oil pipelines is quite remarkable. My local NPR station now regularly covers the Addison Natural Gas Project in Vermont, whose proposed pipeline has inspired a new kind of civil disobedience: the knit-in. The *Wall Street Journal* published an article in 2014 entitled "Protests Slow Pipeline Projects across U.S, Canada." Pushed onto the national stage, pipelines are suddenly common fixtures in our contemporary politics.

So much of this protest has gathered—indeed, overwhelmed—the technical opening provided in the environmental impact assessment of pipelines. Such dissent does not abide by the routine technicalities of environmental impact assessment so much as seize upon its founding promise. The sentiment at these otherwise staid hearings is clear: the impact of fossil fuels grossly exceeds the capacities of life on this planet and must be stopped if we have any chance of survival. By and large, those officials charged with overseeing the process don't quite know what to do with these comments. But in the deluge of protesting

FIGURE 9. 2014 Peoples Climate March in New York City. Photo by author.

comments that overflow the technical process, a new space of political confrontation is opened up where none existed before.

While popular accounts of this sudden interest in energy infrastructure often hang the story on the Keystone XL, there might actually be a much bigger change going on here. Throughout the 2010s and in ways often unnoticed in the United States, oil pipelines around the world have become unexpected battlegrounds in wider environmental disputes. The OCP Pipeline in Ecuador has been a major flashpoint for campaigns to save the rain forest and protect the Indigenous peoples living there. The BTC Pipeline linking the crude reserves of the Caspian Sea to Europe in a way that bypasses Russia became a rallying point for local environmental groups like Green Alternative in Georgia and international groups like Platform UK in London (whose members authored a travelogue of the pipeline, published by Verso in 2012). Although its history is linked to the Keystone XL, the proposed Northern Gateway Pipeline in British Columbia has helped to reactivate distinct genealogies of Indigenous discontent in Canada, working to catalyze the Idle No More movement. So did the water protectors along the Dakota Access Pipeline, which Nick Estes (2019) has so movingly written about. Perhaps the controversy over the Keystone XL is merely the US version of this wider realization of petro-networks.

Marching down Sixth Avenue during the Peoples Climate March, it was striking how present oil pipelines have become. And yet they are present in particular ways. In so much of the spirited antipathy directed at the Keystone XL on that September day, the physical pipeline (and the communities it touches) seemed to matter less than the planetary crisis the Keystone XL has been asked to represent. While Nebraska offers an interesting exception, global warming trumps local context in orienting outrage around the Keystone XL. For many activists I spoke with at the protest, the Keystone XL was not so much a 1,179-mile shortcut in an existing pipeline system as it was the front line in the urgent battle against climate change. Several I spoke with could not even identify the states the pipeline would pass through.

It is notable that the express goal in this rising climate activism around pipelines is not to hijack the physical infrastructure in order to make outsized social demands, like the coal strikers of old. The aim is not to occupy key energy chokepoints but to trip up the tidy image of fossil fuel infrastructure as "safe, silent, and unseen," to borrow a popular industry description of pipelines. In the Peoples Climate March, protesters creatively entangled oil pipelines in narrow conduits of profit, broad patterns of destruction, and the looming possibility of a foreclosed future. The emerging mantra might be: "Don't seize control, seize the implications." In all of this, oil pipelines are being confronted more as ecologies of harm than as buried metal tubes. While the relations of consequence being pinned on petro-infrastructure seem to exceed the communities immediately adjacent to them—as evidenced in both the ambivalence of many towns along the Keystone route and the absence of those communities in climate activism—such relations have nonetheless given these material networks a new scale of transparency. Oil pipelines have become, well, visible. And in that rising visibility, the negative ecologies of fossil fuels are being opened up to new forms of accountability and refusal.

CHAPTER 5

Petrochemical Fallout

PFOA, I'm told, is the slipperiest chemical in existence. Nothing sticks to it, a peculiar quality that found profitable application within the manufacture of plastics. A white, waxy powder first engineered in the 1940s, PFOA helped press Teflon into waterproof fabrics, nonstick kitchenware, and thousands of other consumer products before being washed away without a thought. As shockingly large amounts of this synthetic petrochemical were dumped into the environment, such slipperiness was cast in a more disconcerting light. PFOA resists the forces of decay unfazed; as designed, PFOA is impervious to the ecological jackhammers of microbes, sunlight, heat, and even time. It's also heedless of reified boundaries, slipping through physical bodies, national borders, and earthly mediums like water and air with remarkable ease. PFOA also evaded regulatory scrutiny for decades despite growing evidence of its toxicity.

Yet it turns out that PFOA does get tangled up in some things, like organic matter in soils, activated carbon filtration systems, protein receptors in humans, and a growing number of class action lawsuits. In the past few years, a handful of advocates and lawyers have worked to pull the properties of PFOA into political legibility. Some seventy-five years after PFOA first entered plastics manufacturing, it may no longer be feasible to remove PFOA from the environment. Yet these groups nonetheless labor to call the negative properties and negligent profiteering of PFOA to account and to minimize harm going forward.

Much of this work centers on firming up the science of PFOA, whether by inscribing it with a more exacting molecular identity or by standardizing detection methods and exposure thresholds. These scientific techniques pull PFOA into something tractable, yet in so doing they often lose sight of the bruising confrontations that first brought this petrochemical into public view and glide over the practical demands of impacted communities. PFOA is becoming factual, yet these facts all too often stand at a safe distance from the destructive logics and wounded lives that first called them into being. What forms of understanding, I wonder, might assail those logics and assist these lives?

Although the engineered properties of PFOA are at the center of this chapter, I neither take them up as metaphor nor reduce them to mere chemistry. Rather, I want to understand how select injuries, tactics, and uncertainties take shape around the negative ecologies of PFOA and, in turn, how its negative ecologies materialize new fields of effective knowledge, protest, and culpability. It's the dialectical tension between these two orientations—the open-ended nature of the first and the efforts of the latter to pull PFOA into something more solid and prosecutable—that work to define PFOA contamination today. While exposure may be planetary, the experience of PFOA contamination remains rooted in communities adjacent to plastics manufacturing hubs in the United States and Europe. I happen to live in one of these communities, and I've spent the better part of the past five years working with residents struggling to understand and respond to the discovery of PFOA in our drinking water. In this work, I've often wondered: How might ethnography make PFOA fit within the quite reasonable demands of impacted communities while keeping an eye on all that doesn't yet fit within our present scales of justice?

Today PFOA is found just about everywhere we have thought to look for it: in polar bears and penguins, in snow and rain, in deep aquifers and high mountains, and in just about every form of cellular life on Earth. And now corporate defense attorneys for the petrochemical industry are hard at work nominating PFOA to the welcoming committee of a future of total contamination. It's a future they cast as inevitable, surprisingly democratic, and without any liable author. This corporate argument finds a curious echo in rising currents of anthropological theory. Ethnographic projects in ontology, posthumanism, and queer theory have seized upon the lexicon of toxicity to disavow the political present and prefigure a more radical future unmoored from historical struggle. Slippery indeed.

FIGURE 10. Plastics factory outside Hoosick Falls, NY. Photo by author.

UNSTABLE GROUND

Through no great foresight, I stumbled into this project at a dinner party in early December 2015. After toasts celebrating the end of term, a colleague in the drama department led me to a quiet corner of the room, then asked, "You study disasters, don't you?" She explained a curious addendum she had found in a recent water bill from the Village of Hoosick Falls. Signed by the mayor, the letter noted that the "element perfluorooctanoic acid (PFOA)" had been detected in the wells supplying the public water system, but assured residents the drinking water complied with all federal and state regulations and the levels detected were well below any threshold of concern. "Do you know anything about this?" my colleague asked. When I demurred that I was "not that kind of scientist," she insisted. "Will you please look into this for me?"

A few days later I found myself in Hoosick Falls with three students, not quite sure of what we were looking for and even more unsure of how we'd know it if we found it. A picturesque place, downtown Hoosick Falls is nestled in a wide bend of the Hoosic River where mountain waters tumble over one last rocky ledge before turning across the flat farmland that slope toward the Hudson. The Hoosick Falls Little League

Complex marks the spot where the river swings out and then begins its embracing turn around the town. Just behind left field stands Supply Well #7, a trunk-like steel tube rising a dozen feet from the ground, boxed in with chain link. With only a low hum indicating its heft, the well is capable of pumping nearly a million gallons of drinking water a day from the aquifer below and serves roughly forty-five hundred residents. On that crisp December day, the hum of the well mixed with the dull rumble of the Saint-Gobain High Performance Plastics plant, its roofline rising over the trees some four hundred yards away. In a few days, Brendan Lyons of the Albany *Times Union* would report that Saint-Gobain employees routinely dumped chemicals in the forest separating the factory from the ball field, and the plant had been aware of a plume of carcinogenic chemicals in the groundwater beneath its plant for years but said nothing. But standing by the well on that December day, it was hard to tell what exactly was out of place. A few stops later, we pulled up at the Tops grocery store, where pallets of bottled water had been unloaded, filling the middle of the frozen aisle with boxes of one-gallon water jugs stacked four high.

At the village board meeting that month, as in most previous months, many of the three rows of chairs for residents were empty. Alongside changes to the snow plowing routes, sidewalk improvements, and zoning concerns for a converted apartment building, the last ten minutes were set aside for "the water issue." In a month, these meetings would become major media events packed with television crews; hundreds of angry residents; and huddled teams of city, county, and state officials. The meetings would soon be devoted entirely to PFOA and moved to the basketball court inside the old armory to make room for the van loads of television media. But for now, "the water issue" was one planning issue among a dozen others. The mayor offered a brief update from state agencies on the need for further testing but emphasized the need for calm. The numbers were shared—"642 ppt was detected in Supply Well #7"—but nobody seemed certain of what they meant. That is, nobody except the handful of residents who had begun educating themselves about PFOA. "We all know we've got a problem. Our aquifer is poisoned. What are you going to do about it?" The village board responded to these challenges with polite recusal, deflecting the urgency of such demands with mild-mannered calls for more research and waiting on advice from New York State. Smiling broadly at those gathered, the mayor closed the meeting by noting how "very pleased" he was with progress on the water situation. "It's a very positive situation."

I start with these scenes to emphasize the difficulty of sorting out the discovery of toxicity, not just for the recently arrived ethnographer, but also for residents, many of whom later recalled their initial inability to figure out what was going on. "We just didn't know." "I didn't know who to trust." "I couldn't tell if this was serious or not." The discovery of PFOA was not a straightforward event, and I think it's important to recall the long-standing suspicions, the uneven texture of surfacing questions, and the long shadows they cast (Shapiro 2015; Ahmann 2018). The discovery wasn't a rupture, like a rope snapping under an unbearable weight. Nor was it an awakening, the world drained of its color and left stark and obvious. Despite ongoing efforts of many journalists and advocacy groups to tell this story as one of innocence lost, it wasn't that.[1] It was more of a slow reckoning, nearly impossible to fully face up to. I could never figure out what kind of story I was seeing, let alone what one I should tell. But in my work, a number of encounters left deep impressions. It is from the general experience of uncertainty and gathering constellation of insights that this chapter proceeds. In this, I am interested in how anthropology can attend to unruly fields of indeterminacy that haunt negative ecologies without resigning itself to description without outrage, ethnography without critique, theory without justice.

BECOMING A PROBLEM

Looking back, PFOA was always there. Those summer evenings when a light blue fog drifted across the golf course, and members of the country club quickly moved indoors to finish their meals. Those crisp winter mornings when farmers woke to find their fields painted in blueish hue. There were the recurrent migraines and bloody noses among those living in the new development on the ridge just above the plant. "Some days, I couldn't even go outside," more than one resident told me. Workers called it the "Teflon Flu," an onset of aches and pains after inhaling too deeply while loading the mixers or forgetting to change clothes after getting it on you. Sometimes, of course, you just came down with it for no good reason, other than that you worked at the plant. An electrician told me he dreaded getting contract work in the plant: the pay was great, but something stuck with him when he left, something he couldn't shake for days. A parent told me how the company used to donate industrial barrels for apple bobbing at the community's annual Halloween party. The faint marking of "PFOA" was still visible on the barrels. A mother spoke of the nightmares that wracked her sleep on nights when she could smell the plant emissions. "The ceiling was alive, and it was dripping down and dissolving everything. I could smell it." "In the summer," another resident explained, "you had to remember to close your windows in the evenings. That's when they fired up the stacks." The nights, I heard again and again, smelled of burning plastic.

Afterward—after everyone knew—the long living with it took on a disconcerting question. What hadn't we done? The everyday avoidance of certain smells, certain confrontations, or certain explanations was recast as outrage and complicity, all at once. "We always just had a lot of cancers," one doctor told me, himself recently diagnosed with cancer. "No one really thought about it all that much." So how did the haunting presence of petrochemicals gain political definition as contamination? In the past decade, two new fields of knowledge—the secrets of the corporate archive and the granular metrics of analytical chemistry—were opened in a manner that began to cloth long-standing suspicions of residents in the validation of fact.

Opening the Corporate Archive

Almost from the moment they started producing PFOA in the 1950s, manufacturers had serious concerns about the toxicity of PFOA.[2] As early as 1961, DuPont's head of toxicology said PFOA was likely toxic and should "be handled with extreme care." Over the next decade numerous experiments with animals demonstrated the risks of PFOA. When they didn't all die, these animals developed a host of health problems. As 3M summarized the results: "Certainly more toxic than anticipated." The implications were clear, and both DuPont and 3M started monitoring the health of their workers. Both companies quickly learned that just about any worker who interacted with PFOA, whether in the production line or in the laboratory, had accumulated and retained PFOA in their blood. As the companies sought out an unexposed population to compare the health of their workers with, they encountered a problem: all Americans had detectable amounts of PFOA in their blood in 1976, suggesting universal exposure after only twenty-five years of commercial use.[3] Rather than publicizing this disconcerting fact, 3M instead established an acceptable background level for PFOA in human blood (and found its workers were "1,000 times normal"). That same year, 3M and DuPont got together and collectively decided that their growing concerns about PFOA did not warrant disclosure to the EPA.

After experiments documented increased birth defects in rodents exposed to PFOA (and two women in DuPont's Washington Works plant gave birth to severely deformed children), in 1981 DuPont banned women from working near PFOA. 3M removed all women of "childbearing potential" from PFOA production lines, as an internal 3M memo put it, so "they will not be exposed to fluorochemicals that can cause birth defects in rats"(Lerner 2018). As men took over, new concerns took shape. By the 1980s, DuPont had growing evidence that PFOA was linked to testicular cancer, and 3M found a significant uptick in deaths from prostate cancer among PFOA workers. Within a few years, company scientists linked exposure to PFOA to five cancers in animal studies. By 1988, DuPont had internally classified PFOA as a "human carcinogen" while continuing the tell the EPA it was benign (Lerner 2015). In 1997, 3M updated its data sheet on PFOA to read: "CANCER: WARNING: contains a chemical which can cause cancer," citing joint studies by 3M and DuPont from 1983 and 1993 (Lerner 2018). 3M then removed the warning, and a year later 3M executives told the EPA that the company had no data suggesting human health risks associated with PFOA.

Nor were health concerns limited to factory workers. In 1982, a DuPont memo noted "there is obviously great potential for current or future exposure of members of the local community from emissions leaving the Plant perimeter"(Lerner 2015). Two years later, DuPont asked a few employees to visit the towns that line the Ohio River just downstream of the Washington Works Plant and fill jars with their drinking water. Analysis revealed PFOA at levels two to three times DuPont's own internal safety standard for drinking water. After some discussion, DuPont decided to say nothing. That same year, 3M executives instructed employees to "clean computers of all technical data" regarding health concerns about PFOA (Lerner 2018). Requests for further PFOA health studies at DuPont were denied, with the explanation: "Do the study after we're sued" (Lerner 2015).

3M and DuPont's strategy of burying what they knew about PFOA almost worked. But then a corporate lawyer from Cincinnati started sniffing around DuPont's Washington Works plant on a pro bono basis for an old family friend. After DuPont built a landfill just upstream of his farm, Wilbur Tennant watched his cattle start acting deranged and dying rather macabre deaths.[4] Tennant complained, but DuPont controlled the town, and he soon found himself ostracized by the community. Undeterred, he eventually tracked down the grandson of a family friend. Although he typically worked the other side of cases like this, corporate defense attorney Rob Bilott, who had ridden horses on the Tennant farm as a child, was moved by the story and filed suit against DuPont in 1999. In the lead-up to the trial, Bilott stumbled across a DuPont memo about a chemical named PFOA that was delivered to the landfill upstream of the Tennant farm. Unable to find any public information about this chemical, Bilott requested the company records on PFOA. DuPont refused, but eventually a judge required them to comply. The files—some 110,000 pages in all—soon filled his office, and then some. Thumbing through them, Bilott discovered a comprehensive archive of the dangers of PFOA. DuPont quickly settled the case with Tennant and tried unsuccessfully to get a gag order blocking Bilott from disclosing what he had learned. Bilott compiled a nine-hundred-page brief that outlined DuPont's lengthy record of internal concerns about PFOA and delivered it to the EPA in 2001. Bilott's now famous letter called PFOA an "imminent and substantial threat to health or the environment" and demanded the EPA begin the process of regulating PFOA. The corporate archive had gone public.

Due to these revelations, DuPont reached a $16.5 million settlement with the EPA in 2005 (at the time, the largest civil penalty the EPA had levied) for not disclosing its internal alarms about PFOA. Due to this lawsuit, in 2006 the EPA facilitated an agreement to a voluntary phaseout of PFOA from manufacturing streams in the United States by 2015 (worldwide production of PFOA has stayed level as plastics production simply moved to China).[5] A few months before the PFOA phaseout was complete, the EPA issued an "emerging contaminant" fact sheet that summarized what the agency had learned from DuPont and 3M about the health risks of PFOA and, based on those disclosures, issued advisory guidance levels for acceptable levels of PFOA in drinking water (set at 400 ppt). Some sixty years after PFOA entered plastics production in the United States and some fifty years after 3M and DuPont realized that exposure to PFOA had detrimental effects on human health, the EPA issued its first public recognition of the dangers of PFOA, in March 2014.

New Analytical Sensitivity

Measuring toxicity on the scale of one part in a trillion (ppt)—that is, an analytic sensitivity equivalent to finding one person on two earths—is a recent achievement. It comes with a laundry list of caveats but largely rests on the reinvention of a century-old device: mass spectrometry, or the mechanical art of weighing atoms. In the past few decades, liquid mass spectrometry (LC-MS) has ushered in a minor revolution in the granularity of environmental science and policy. The mass spectrometer uses what is called "targeted compound analysis."[6] That is, the incredibly sharp resolution of LC-MS is relative to the precision of the standards used. Or, as an analytic chemist explained it to me, "With the Liquid Mass Spectrometer, I can only see what I'm looking for." LC-MS does not measure the concentration of a chemical in a sample directly; it measures a sample in relation to a standard concentration.

Perhaps no chemical has exemplified the promise and predicament of LC-MS analysis as forcefully as PFOA. While worker cohort studies demonstrated clear links between PFOA and a host of adverse health impacts inside plastics factories, the problem of exposure outside the factory was largely invisible until LC-MS made parts per trillion visible. "It's only with the LC-MS that we've began to see PFOA in the environment," one analytic chemist told me. With LC-MS, as a leading environmental engineer further noted, "It's only in the last five years have we gained the ability to reliably quantify PFOA in water and soil." (And, I might add, in our bodies.) Why? "PFOA standards of the accuracy required for LC-MS have only recently become commercially available," an analytic chemist working on PFOA explained to me. Until 2003 DuPont and 3M controlled all the labs that could measure and monitor PFOA, in part by claiming the LC-MS standards they had developed for PFOA were a proprietary method.[7] It took a series of lawsuits to finally make those standards available to commercial labs and academic researchers. Although its role remains largely uncredited, the analytic sensitivity of LC-MS was key to allowing epidemiological and toxicological experiments to begin asking new questions about the health impacts of PFOA, especially in the form most people are exposed to it: at extremely diluted levels in drinking water.

Advances in LC-MS opened new ways of grasping the unprecedented ecological properties of PFOA. PFOA is extraordinarily durable. Comprised of a synthetic carbon-fluorine bond—described to me as the "Hercules of chemical bonds" or "molecular rebar"—PFOA is nearly indestructible. Unlike other persistent organic pollutants (POPs), as one environmental engineer explained, "There is no known natural degradation or decomposition process for PFOA." Whether exposed to extreme high or low temperatures, the radiating energy of the sun, the appetites of microbes and fungi, or even caustic chemicals, PFOA does not break down. Chemically stable on the order of centuries (a lifespan, I was told, that is based not on empirical data but on the fact that most existing models for contaminant breakdown don't figure timescales beyond a century), PFOA is, as one regulator told me, "redefining the concept of environmental persistence." "This stuff exists on geological timescales," another environmental engineer quipped. For most of the past century, the profound inertness of PFOA was taken as evidence of its benign nature. Toxicity was widely understood to be in accordance with the bioreactivity of a chemical. Today, the inertness of PFOA is coming into new focus, not as evidence of its harmlessness but as an intricate vehicle of a new regime of toxicity.[8]

PFOA has forced the realization, as one EPA scientist explained to me, "that even if a contaminant doesn't react with anything, it can have toxic effects." Bioaccumulating in humans, PFOA has a half-life of about three to five years, for reasons that are still poorly understood. ("3–5 years is an eternity for a chemical to stay in the body," one EPA toxicologist told me.) Most striking, this toxicity is evident at extraordinarily low levels of exposure. Indeed, research into PFOA is adding a whole new metric to toxicology: parts per trillion. ("At that level, you're basically count- ing individual molecules," one environmental scientist told me.) Today, extremely low levels of exposure to PFOA—primarily through drink- ing water—have been strongly correlated to developmental disorders, immune deficiencies, reproductive abnormalities, thyroid disease, pros- trate cancer, testicular cancer, and kidney cancer. Moreover, there is some suspicion that PFOA does not follow the "dose-response" hypothesis— the more you get, the sicker you get—that underlies contemporary toxi- cology. Instead of a clear dose-response curve, study after study of PFOA has demonstrated a clear statistical scattershot of adverse health impacts after a population is exposed to PFOA (often at barely detectable levels). The leading EPA toxicologist working on PFOA told me, "There is no threshold under which there are no effects. There was only low effects." In this, PFOA is not only overwhelming the dominant paradigm of toxi- cology, it is also breaking the backbone of toxic regulations: thresholds.

These groundbreaking concerns—the properties of PFOA brought into focus by LC-MS analysis—are not restricted to the pursuit of envi- ronmental responsibility. The minuscule measurement of PFOA in water, soil, and blood has also opened up lucrative frontiers for analytical chem- istry. An annual conference for environmental scientists hosted its first panel on PFOA in 2016. The year before, an organizer told me, they had tried to put together a panel, but there was no interest. One year later, it was standing room only. Several hundred environmental consultants, municipal and state officials, and a handful of industry representatives packed the room, with folks standing in the back and squatting in the aisles after seating filled up. "PFOA," the organizer said, leaning over to me, "is blowing up." Michael, the senior chemist at a commercial lab, began his presentation on analytic techniques for PFOA. "It's a huge growth area for us," he said. And with all the lawsuits in the works, the market was exploding. "You can make a fortune on this." His second slide was a hodgepodge of newspaper clippings about communities impacted by PFOA in Parkersburg, West Virginia; Hoosick Falls, New York; and Bennington, Vermont. "Obligatory headlines of why we care," he said, to light laughter from the crowd.

Citizen Outrage

Michael Hickey certainly stands out. Raised in Hoosick Falls, New York, Michael is well known for wearing bright pink oxford button-downs in a town that lives in the faded palate of Carhartts and blue jeans. When his father passed away from kidney cancer in 2014, Michael was devastated. In his grief, Michael realized he had watched far too many of his classmates be diagnosed with cancer in their twenties and thirties. As the town doctor later told me, "There just always seemed to be a lot of cancers in this town." Although Michael's father never smoked or drank, he worked in the Saint-Gobain High-Performance Plastics factory right up until his diagnosis. Late one night Michael googled "Plastics" and "Cancer" and started reading. He quickly narrowed in on PFOA and started compiling a number of articles (many just published that year by the EPA) that linked PFOA to various cancers and other health impacts. The modest plastics plants that dot the Bennington/Hoosick Falls region—which renamed itself "Teflon Town" in the 1970s—emitted PFOA by the ton annually with no regard for what happened next. In Hoosick Falls, the Saint-Gobain plant sits about three hundred meters from the well that supplies the town's drinking water. Michael wondered if PFOA had contaminated that drinking water. "I started staying up a couple of nights a week for 3 months reading up on it," he explained. Alarmed at what he was learning but unsure of its validity, he took his pile of articles and notes to his doctor and asked him to take a look. The doctor thought Michael might be on to something. Encouraged, Michael took his findings to the village board in early 2014 and suggested the village test its drinking water for PFOA. "I thought it'd be a no brainer," Michael said.

Instead it took two years of tireless work by a growing coalition of outraged residents to bring the question into public light.[9] Hickey convinced a commercial laboratory that had just developed an LC-MS analysis for PFOA to run samples from Hoosick Falls.[10] When the town refrained from providing samples from the well, Hickey took samples from his mother's house and the dollar store, among other places. It took this rogue sampling to confirm Hickey's suspicions: alarming levels of PFOA were in the town's drinking water. Then it took an indomitable will to stand against the mayor, the county health department, and finally the New York State Department of Health, all of whom continued to tell residents there was nothing to worry about long after evidence to the contrary was overwhelming. As we now know, PFOA is surprisingly mobile, is unfazed by any natural degradation process—environmental advocates now call PFOA a "forever chemical" for its sheer indestructability—and when consumed, quickly accumulates in the human body, where it is strongly linked to a host of cancers and other ailments. These disconcerting properties of PFOA have been abundantly clear to the plastics industry for decades and to the EPA since at least 2005. Until quite recently, however, it was not clarity but confusion that met communities like Hoosick Falls when they discovered PFOA in their drinking water.

At one public meeting in Hoosick Falls in December 2015, a citizens' group—a collection of furious mothers joined by local doctors, lawyers, and bankers from the town—manned a table in the back of the auditorium. There they handed out a recently published EPA fact sheet on PFOA that, in highly technical language, summarized growing concerns over the "toxicity, mobility, and bioaccumulation potential of PFOA" at the levels found in the town's drinking water. At the front of the room, New York State Department of Health officials gave a presentation that explained that while PFOA had been detected in the public water of Hoosick Falls, "health effects are not expected to occur from the normal use of the water." A mother interrupted, "We all know we've got a problem. Our aquifer is poisoned. What are you going to do about it?" The Mayor told everyone within earshot that drinking the water was "a personal choice," and while he understood why some people were choosing not to drink the water, he would continue to drink it.

The work of Michael Hickey and other residents suddenly pricked the "everyday praxis of not noticing" in these communities and drew the long-standing chemical milieu of plastics manufacturing into the density of a moral event (Ahmann 2018: 145). The revealed secrets of the corporate archive and the new granularity of LC-MS helped elevate perennial suspicions "into events that stir ethical consideration and potential intervention" (Shapiro 2015: 369). These emerging fields of knowledge provided residents with the solid ground to disavow cognitive and bureaucratic investments in "toxic uncertainty" (Auyero and Swistun 2009).[11] Residents came to demand answers to questions that had long hung in the air and began organizing for justice in their community.

A Public Problem

This is how PFOA became a public problem. The three perspectives discussed previously—the opening of the corporate archive at 3M and DuPont, new analytical sensitivity with the LC-MS, and the welling up of citizen outrage—aligned and together worked to stabilize PFOA for long enough to start doing something about it. For frontline communities, this convergence of legal, scientific, and local understandings of the negative ecologies of PFOA transformed a suspicion that had long hung in the air into a crisis with the firmness of fact. Together, this turned PFOA into something you could situate suffering in relation to, you could organize against, and you could make demands from. It turned PFOA into a problem you could see scientifically and confront politically. This potent convergence—and the reckoning it made possible as the potential cost of cleaning up PFOA and paying for the medical bills of everyone impacted grossly exceeded the market valuation of the handful of companies responsible—pushed the petrochemical industry into a novel defense strategy. With their archives pried opened, the traces of their products now found everywhere (and everywhere causing harm), and communities rising in outrage, corporations like 3M and DuPont have shifted gears. Arguing that we've passed the point of no return, lawyers and lobbyists for the petrochemical industry now reject site-specific remediation (and liability) as an antiquated response to the planetary futures chemicals like PFOA inaugurate. Total contamination is the starting point of the contemporary condition. This corporate stance finds a curious echo in new theories of toxicity in anthropology.

Organizing against Contamination

Keith built his modest house on the ridge above the plastics plant. A skilled carpenter, he sometimes spoke of his home as a kind of college degree, an investment in his future. After dropping out of high school and working in construction for years, it was the first real project he had taken on himself, and the skills he had honed proved themselves as he became a noted housebuilder in the area. In the past few years, his home has been cast in a different light. PFOA from the Saint-Gobain plant infiltrated the groundwater in North Bennington, Vermont, and in 2016 many residents discovered high levels of toxins like PFOA in their residential wells. The PFOA levels in Keith's well were one hundred times over what the state of Vermont deemed safe. It was a real puzzle though, as the houses around him had barely detectable levels of PFOA. Every time I'd come for a sample, he'd follow me to the basement to chat. "My boys still won't drink the water," he told me one afternoon. "I tried to explain it to them, but they just won't do it. Won't even use the water to brush their teeth. They are still afraid of it." He didn't find their fear silly, he said; he understood it. But he wasn't sure how to square it with the filtration system Saint-Gobain had installed in his house. So he keeps buying his children bottled water, on his own dime.

When I first met Emily she had a hand-painted sign that proclaimed "Cloud Nine" staked on a knoll overlooking her driveway. A few weeks later, the sign leaned up against the shed. A few months later it was replaced with a "For Sale" sign. "It's no longer my house," Emily said. "It's theirs." She pointed at Taconic Plastics, just down the road. "Once they poisoned my water, they took away my home."

Emily worked three jobs until she could pull her children out of a decrepit two-room trailer and into her dream house: as she described it, "a 3 bedroom, 2.8 acres, American Dream. Did it before I was thirty, and while I was single. I loved it." In 2016, Emily was informed that PFOA had been detected in her well at levels over thirty times the federal health guidance level for short-term exposure. She was devastated. State officials asked her to wait patiently while they worked something out with the company. She didn't, and as she tried to bring attention to the issue, friends rebuffed her. The former town supervisor cornered her: "Do you really want to cost 200 people their jobs over this?" She prevailed, and against tremendous headwinds forced the issue into the light of day, much to the embarrassment of company leaders and state agencies who had been sitting the problem for a decade without telling anyone. It's a story she has recounted many times for television crews and legislative hearings since then, and when I arrived with students to sample her water, she'd always tell it. One morning, she flashed a grin after relating her story. She said she had a surprise to share: "I'm pregnant." As I offered my hesitant congratulations, she interrupted me: "Does anyone need any breast milk? Cause I don't. My blood levels are too high. I'm not going to pass these chemicals on to my baby."

Each of these towns lived with plastics manufacturing for decades— in the 1970s my adopted hometown of Bennington rebranded itself "Teflon Town" in celebration of the new hub of plastics manufacturing that was sprouting up in the old mills that dot the region (Therrien 2017)—and had lived with the contamination of their lives for just as long. Long registered in an indigo-tinged fog on winter mornings, chronic nose bleeds, the acrid smell of plastic burning in the summer, headaches, tap water foaming as if already soapy, and cancers among family and friends, PFOA contamination was not exactly a revelation to these communities. Memories of "before" were not exactly colored with innocence.

For many residents, the shape of injustice gathered into felt form around the two remaining social safety nets in rural America: family and home. In these working-class communities, family and home are often talked about more in terms of reciprocity than gain: folks pour their labor into their families and homes with some hope they will eventually return the favor with care, meaning, and stability in regions otherwise bereft. PFOA smuggled profound harm into the two vestiges of well-being left in these downwardly mobile communities. And that's where long-tolerated risk snapped into welled-up fury over PFOA contamination. Michael Hickey later reflected, "I'm not a doctor or a lawyer or even an environmentalist. But I knew something wasn't right. I started as a heart-broken son and quickly turned into a scared father." Residents organized as mothers and fathers. They protested as homeowners. At public meetings, residents explained the impact in terms of children now carrying a lifetime of medical uncertainty and their meager life savings being wiped out in collapsing real estate prices. These two ledgers of loss formed the basis of how residents drew long-standing exposure into demands for justice.

As they organized, residents worked together to minimize exposure to PFOA going forward, to secure medical support adequate to the lifetime of worry their families now carried, and to advocate for robust regulatory protections from toxins. Contamination was not total, and what justice remained was found in efforts to contain toxicity and find redress for injuries already underway. Much of this work involved pulling PFOA into the legibility of the environment, whether by demanding lower thresholds for levels in drinking water or demanding impact assessments of knee-jerk efforts to dispose of PFOA stockpiles. As I became more and more involved in these efforts, I found myself aligned with the environment, arguing for thresholds and impact assessments, arguing with disembodied facts, or even inhabiting the bureaucratic procedures of the environment with the hopes of slightly enlarging their reach. While I remain convinced of the ultimate complicity of the environment and the oil industry, this work made me newly appreciative of how environmental protections can and do provide a means of holding back the worst. It also taught me how little such realities figure into the growing theoretical stature of contamination in anthropology today.

THEORIZING WITH CONTAMINATION

"Everyone carries a history of contamination; purity is not an option," writes Anna Tsing (2017: 27). Today, many anthropologists are moving from documenting the underlying hybridity of the modern world to aligning ethnographic inquiry with problems that actively propel transgressions beyond modernity. Perhaps no other topic tracks this unfolding shift like toxics. In these troubling times, a growing number of anthropologists have found renewed theoretical optimism in the chemical capacity of contamination to scramble modernist strictures and inject experimental hybrids into our now unprecedented future. Describing how rusty chemical weapons in Panama author new multispecies assemblages, Eben Kirskey (2017: 1) calls for anthropology to learn how to "experience the dangerous pleasures of intoxication." Elizabeth Povinelli (2017: 509) imagines our bodies, impinged upon by rising seas and feral toxicity, "are stew pots cooking up a new form of posthuman politics." Toxics physically upend purified epistemologies and their staid political forms and in so doing open the empirical possibility for critical scholarship to root itself in transgressed boundaries, denaturalized dualisms, and the gathered anticipation of worlds to come.[12] Contamination is a fait accompli, and ethnographic writing can help spur this insurgent truth by seizing upon its world-making possibilities.

"Toxic environments are animating transgressions," writes Eben Kirskey (2017: 1). Many prominent theoretical voices in anthropology today are converging on contamination as a physical rupture with the epistemic habits that undergird colonial modernity, as a kind of revolutionary release from the categorical reason that got us into this mess. Contamination instigates hybrid experiments and planetary futures that may finally break the death grip capitalism and state power have on our present. All too often, bemoaning the harms of toxins ascribes yet another deficit to marginalized groups with scholarship that further "surveils and pathologizes already dispossessed communities," as Michelle Murphy (2017: 496) warns. Moreover, environmental justice scholarship around toxicity often pivots on "a hopeful relation to the state" that can paradoxically work to morally legitimate the very agencies that permitted contamination in the first place (Murphy 2008: 699). Against such damaged complicity, anthropology should embrace the insurgent possibilities of contamination. Kirskey (2017: 2) calls for "toxic methods" in anthropology more attentive to the world-making capacities of chemically "altered abilities and subjectivities" in contaminated worlds. Nading (2020) advances a programmatic call for "anthropological toxic worlding."

Much of this work in ethnography explicitly draws inspiration from a branch of queer theory that is convinced of the emancipatory possibilities of toxics to disrupt structural binaries and secure a more radical future (see Di Chiro 2010; Weiss 2016).[13] Antke Engel and Renate Lorenz (2013: 5,1 0) embrace the widening reach of toxicity "as a means of queering subjectivity and sociality" and "destroying the system from within." Morgan Holmes (2000: 103) discusses how petrochemicals and synthetic hormones "threaten the hegemony of heterosexuality," concluding that toxic contamination "is a quite promising kind of troublemaking." Toxicity, write Malin Ah-King and Eve Hayward (2013: 7), now outpaces "social or political movements" in advancing queer politics by way of "metabolizing pollutants, xenotransplanting toxicants, and intravenous banes." Reveling in how "toxicity releases life from an absolute need to protect it," Mel Chen (2011: 279) asks if we might recenter our research, ethics, and politics on "the queer productivity of toxicity and toxins" and seize upon the world to come. Thumbing through the harms of toxicity today—breast cancer, prostate cancer, lowered fertility, intersex characteristics, and deformed children—Anne Pollock (2016: 183) asks why "no one is celebrating the queer here."[14]

UNIVERSAL CONTAMINATION

"Toxicity is now a planetary force," writes Joseph Masco (2015: 144). Michelle Murphy (2008) outlines the "chemical regimes of living" whereby the pathways of industrial emissions, agricultural pesticides, and synthetic hormones now alter the molecular composition of life worldwide. From the unbound fallout of nuclear weapons to the lived imprint of petrochemical prosperity, contamination has gone global and should be acknowledged as such. In the 1960s and 1970s, the hemispheric afterlives of DDT and strontium 90 proved instrumental in provoking a new age of environmental reason and responsibility. Today, wider arrays of contamination, so many of them tied up with fossil fuels and petrochemicals, convincingly mark out our epochal lurch into planetary and cellular instability. The near universal imprint of radioactive waste, automobile exhaust, plastics, farm runoff, industrial smog, acid rain, and pollution all work to index the geological coordinates, historical rupture, and embodied precarity of our impending future. Toxicity, writes Gabrielle Hecht (2018: 1), is one of the "foundational categories" of the Anthropocene. Indeed, as one recent volume suggests (Tsing et al. 2017), toxicity may be the charter entanglement of our planetary crisis

Across wide bodies of scholarship, today's unbridled toxicity brings the worsening planetary condition into crisp empirical and conceptual focus. Yet so often the theoretical novelty of the contaminated planet is found in its ability to highlight just how feeble modernist forms of sovereignty have become when placed on a more planetary stage. Radioactive fallout, writes Joseph Masco (2015: 144), unleashed "invisible injuries" whose legibility was deeply reliant on national security infrastructure, yet whose affective texture and ecological reach always exceeded the operations of that increasingly dated political form (indeed, a state whose very "datedness" was in no small measure the result of planetary toxicity). Describing how pollution and other exposures defy the reason and reach of the modern state, Bruno Latour finds a new politics beckoning from our rising collective insecurity. "The new universality consists in feeling that the ground is in the process of giving way," Latour (2018: 9) writes, before pondering how we might regroup in the freefall. Describing how contamination haunts landscapes but perhaps more hesitant about universal invocations, Anna Tsing, Heather Swanson, Elaine Gan, and Nils Bubandt (2017) describe how feral toxicity might help us learn to inhabit the ruins of modernist progress with posthuman humility and multispecies solidarity. The open-ended spread of contamination today demonstrates the fatal shortcomings of liberal governance and market progress while suggestively tracing out the basis of a more encompassing politics. In this, perhaps the future of contamination is replacing the history of empire as the main stage for theorizing the built-in blind spots of liberalism. The planetary fact of toxicity is both a devastating critique of our impoverished institutional capacities and an emboldened road map for what might come next. While hugely generative for social theory, this turn toward what exactly our contaminated planet heralds so often begins by first turning its back to the jurisdictions required to prosecute the injuries of toxicity today.[15] It is also a theoretical stance that the petrochemical industry is finding advantageous.

After they became aware of PFOA's unprecedented toxicity, 3M and DuPont continued to manufacture more than enough PFAS to poison the drinking water of every single resident of North America. Refusing to disclose the immense dangers these chemicals posed, PFOA was sold to plastics plants, carpet and shoe factories, and even oil and gas drilling sites all across the United States, where it was routinely discarded by the ton into the air and water. Some industries even endorsed the distribution of PFOA-laden waste to local farmers as a soil supplement.

Each time the question of containing PFOA came into view, 3M, DuPont, and now Chemours launched a perfluorinated blitzkrieg in the opposite direction. They flooded the zone. And looking back, a rather demented product defense strategy becomes apparent: total contamination. Rather than controlling PFOA contamination, the petrochemical industry universalized it. Today, PFOA is everywhere. And now that very ubiquity is now being voiced as an excuse for not doing anything about it. How do you remedy a toxicity so endemic to our lives that it's become impossible to rewind?

In March 2018, Saint-Gobain submitted the final report of its investigation into PFOA contamination in southern Vermont. Weighing in at a hefty 7,377 pages and claiming to be the final word on the matter, the report offered a deeply technical (and deeply cynical) definition of PFOA contamination. Many local news organizations ignored the report, and when I asked why, they pointed to their lack of capacity to navigate the ocean of field logs, laboratory reports, and technical details in the 7,000 pages. Yet the report was quite consequential and, if uncontested, it would severely constrain the scope of both remediation and responsibility. The report showed PFOA emissions from the factory were extremely modest and geographically contained: emissions fell back to the ground in the neighborhood immediately around the plant. Yet the report also showed that PFOA contamination was extensive across the entire region and beyond. How did this make sense? Saint-Gobain used the report to argue that the region had high background levels of PFOA through emissions from distant industrial sources and the irresponsible waste disposal practices of local residents. Against these high background levels, the specific contribution of Saint-Gobain

FIGURE 11. Community meeting at Bennington College about PFOA contamination in 2016. Photo by author.

was exceedingly small if not entirely negligible (as would be its liability). PFOA contamination is so extensive that who can really say who is responsible?

On one level, it was a fiendishly clever argument, and state agencies struggled to contest it within the already agreed upon scope of the investigation. Working with students and colleagues, I worked to pull this argument and its significance into public light through op-eds (Bond and Rose 2018). In the past few years, state investigations into PFOA contamination have commenced in nearly every state (and all around the world), and there are hundreds of lawsuits from impacted communities currently underway. Some analysts now worry that liability for PFOA in the United States may equal the market valuation of 3M and DuPont (DePass 2019; Root 2019). If contamination beyond the United States is considered, that liability could easily exceed the market valuation of the petrochemical industry. In response, lobbyists and lawyers for the plastics industry now project PFOA as a preview of universal contamination. It's a reality they cast as inevitable, surprisingly democratic, and without any liable author.

New anthropological questions of toxicity and planetary futures have garnered much theoretical excitement and continue to make key contributions to debates on the Anthropocene. Yet such work struggles to reflect on the strange political bedfellows such a stance may be making. Today petrochemical industry and fossil fuels companies are also investing heavily in the fact of planetary contamination as the starting point of the contemporary world. Turning away from the failures of the past, the generative look into planetary futures can tune out more exacting geographies of liability. Toxicity *is* a planetary issue. But it is one so often profited and fielded, disavowed and inhabited in grossly unequal ways: contamination is "a condition that is shared, but unevenly so, and which divides us as much as binds us" (Murphy 2017: 497). How can anthropology acknowledge unbridled planetary contamination of fossil fuels and petrochemicals while working to hold accountable those who have profited from toxicity? How might ethnography become better attuned to the historical inequities and planetary futures that haunt questions of toxicity today, in full awareness of the political stakes at work in these scales of reckoning?

THE WORLDLY RECOGNITION OF PFOA

It's in your blood. It's in your food. And there's a good chance it's in your drinking water. With PFOA, one toxicologist told me, "We all have body burdens now." In the past few years, various projects have worked to bring the negative ecologies of PFOA into public view. The slipperiness of PFOA is being pulled into the fixity of remediation profits, regulatory science, and struggles for justice. Yet even within these projects, the negative ecologies of PFOA continue to exceed the best solutions of the state.

After the briefest moment of utility, PFOA plagues the earth with roving mobility, immortal toxicity, and an affinity for living things. Exposure to PFOA at exceedingly minuscule levels is strongly linked to developmental disorders, immune dysfunction, male infertility, and a host of cancers. The obstinate toxicity of PFOA was clear to the petrochemical industry from the beginning. Yet for most of the past century, these worrisome effects were hidden away in corporate archives, military secrets, and lackadaisical oversight while the petrochemical industry dumped obscene amounts of these synthetic per- and polyfluoroalkyl substances (PFAS) like PFOA into our air, our water, and our lives.

What choice do we now have, the petrochemical industry sagely observes, but to resign ourselves to a future of irreversible and complete PFAS contamination?

It's a bit like Wall Street asking for a government bailout after bankrupting the entire economy, reasonable only if you forget everything that got us into this mess. Yet the EPA and far too many state agencies are beginning to accept such a corrupt premise. Discovering PFAS in drinking water, in farms, in aquifers, and in all of our bodies, those tasked with protecting public health are throwing up their hands at the sheer ubiquity of the problem. PFAS is becoming too toxic to fail.

PFAS is now everywhere, but this disconcerting fact should not distract us from the petrochemical operations holding the smoking gun. Smoking, in no small part, because they are still emitting PFAS. The omnipresence of PFAS does not lessen the threat they pose to our health, but it does mean we need bolder ways of prosecuting these monumental environmental crimes against humanity. PFAS contamination demonstrates how hamstrung our toxic regulations have become when faced with the deranged profiteering of the petrochemical industry. Surely such shortcomings should prompt a deepening of our commitment to environmental justice and a broadening of our confrontation with fossil fuels.

Instead, EPA and many state agencies appear ready to throw in the towel. EPA's *National PFAS Testing Strategy* bemoans how "impossible" it is for "EPA to expeditiously understand, let alone address, the risks these substances may pose to human health and the environment" (EPA 2021: 3). Overwhelmed by the hordes of PFAS that assail our health today, EPA is going ask the petrochemical industry to study a handful of these chemicals in the hopes that industry-commissioned research will build up an effective case for regulating one chemical at a time. The timeline proposed will take another century (or two) to make its way through the entire family of PFAS, which now number in the thousands. Heeding the corporate siren call of resignation, those charged with protecting our environment take the planetary reality of PFAS contamination as proof of their diminished stature in a world already lost.

Agencies propose natural "background levels" for a synthetic chemical conjured up a mere seventy-five years ago, in effect giving tacit approval for the history of gross negligence that got us here. Agencies spread blame back to residents by listing household items containing trace amounts of PFAS alongside factories that emitted it by the ton annually as equivalent sources of local contamination. Agencies refrain from sampling groundwater near industries suspected of using PFAS, wary of what they might find (and their own complicit delay in coming to the question). Agencies stack science committees assigned with reviewing PFAS toxicity with industry lobbyists while putting up roadblocks for independent academics to participate. Agencies applaud a pyrrhic victory of finally deciding to regulate PFOA and PFOS twenty years after their rampant toxicity was revealed, while the petrochemical industry churns out a witch's brew of other unregulated PFAS chemicals. Agencies endorse incineration as the disposal method for PFAS while acknowledging there is no evidence that combustion destroys these flameproof chemicals. Agencies make grand commitments to keep studying the problem in the hopes of taking action in a decade or so.

The point is clear: by way of regulatory indifference, delay, and now despair, responsibility for the toxicity of forever chemicals is shifting from the corporations who profited from them to the communities who must now live with them.

All is not lost. While PFAS inspires paralysis in environmental agencies, people living on the front lines of this crisis demand action now. Aware there is no going back, rural towns next to military bases, working-class neighborhoods adjacent to plastics factories, and communities of color near incinerators burning PFAS insist we do everything we can right now. They demand an immediate stop to all ongoing releases of PFAS. They demand we compel the industry and the military to start cleaning up sources of PFAS contamination. They demand we ban PFAS as a family of chemicals, not only in the United States but across the world. They demand we pass the PFAS Accountability Act, legislation that insists manufacturers retain liability for all the damage PFSA inflicts after the factory. And they demand we hold polluters fully accountable for the decades of damage they've done. By any means necessary, these communities insist polluters pay for water filtration systems for every single impacted home and business, pay for medical monitoring for the lifetime of worry they now carry, and pay for independently scientific monitoring over the generations that PFAS will haunt their community.

From the beginning, state agencies in New York and Vermont promised to "make everyone whole again." It's a familiar refrain from officials, one that works to set a retrospective baseline before pollution as the technical goal of remediation. For many residents, their longer familiarity with contamination, the injuries they carry forward, and the unique chemical properties of PFOA pointed in a different direction. Many know: there is no going back. It was not nostalgia that drove their protests but securing a better world today. Residents organized themselves around advocacy for clean drinking water (with the recognition that there are no perfect options) and help with medical bills for ailments linked to exposure to PFOA (with the recognition that they will carry a lifetime of risks). Residents' pursuit of practical justice also reoriented their understanding of their place in the world. The largely white, working-class communities of Hoosick Falls and Bennington have hosted mothers from Flint, Michigan; sent care packages to the water protectors at Standing Rock; collaborated with high schoolers from east LA working on drinking water issues; published op-eds in communities around the United States that have discovered PFOA in their water; and reached out to communities around similar plastics plants in India and China. In 2018, the congressional district representing Hoosick Falls flipped from Republican to Democratic, largely on the issue of water protections. Their confrontation with PFOA has keyed them in to the wider struggles against contamination today and demands for justice in the present tense.

CHAPTER 6

The Ecological Mangrove

While I was conducting fieldwork on St. Croix in 2011, the HOVENSA refinery on the island broke down in spectacular fashion. That summer explosions rattled the neighborhoods around the refinery as black smoke draped the verdant landscape in what looked like sooty cloaks. Explosion after explosion was followed by emergency warnings to shelter indoors, and refinery employees in hazmat suits went door to door to skim the oily surface off residential rain catchment basins. I had arrived in St. Croix to study the political economy of oil refining both on this modest US territory and in the wider Caribbean. Yet soon my attention was drawn to the looming sense of ecological dread that rose on the island with these explosions. As I soon learned, these spectacular breakdowns in 2011 stood atop a long history of destruction at the oil refinery, and I was soon swept into the even longer history of Caribbean oil spills in factualizing the magnificent ecology of mangroves. This history put new scientific winds in the sails of a distinctive Caribbean sense of belonging that was able to both fully acknowledge the colonial history of the region and chart a way beyond it. This chapter stays close to how the tentacles of leaky oil refineries assaulted marine life across the Caribbean, and how those negative ecologies imbued mangroves with new prominence in public debates about the constitution of the Caribbean in and against imperialism.

This chapter, then, examines the collision of oil refining and mangroves in the Caribbean during the 1970s as one regional facet of a larger dialectic of fossil fuels and ecology. Such a collision, I argue, offers a striking account of how the agency of the natural world became intelligible within the contemporary world. Beginning with the pivotal but often neglected place of petroleum in the Caribbean, I show how crude oil still requires some explanation in the region. Using the local history of what became the largest refinery in the Western Hemisphere—the mammoth scale of this St. Croix refinery is dwarfed only by its neglect in popular and scholarly accounts of the region—I describe how fossil fuels were introduced to one colonial territory in the Caribbean and the social and environmental consequences of that introduction.[1] Over a dozen export-oriented refineries were built by US oil companies in a similar fashion across the Caribbean between 1950 and 1970. As the United States moved significant portions of its hydrocarbon infrastructure offshore, the region became the world's largest exporter of refined petroleum products in the world, almost all of which went to the United States (United Nations 1980). Between 1950 and 1990, oil refineries became the largest site of capital investment in the Caribbean, a leading source of state revenue, and one of the region's largest employers, especially during the construction boom of refineries in the late 1960s and early 1970s (United Nations 1979; World Bank 1984; Richardson 1992). These events stand at odds with other accounts of what makes the Caribbean a unique and enduring cultural region. As the sugar plantation became the defining image of the Caribbean for critical scholars and national leaders alike, the expanding energy networks of the United States underwrote much of the area's contemporary aspirations.

This petro-economic boom unleashed its own petro-ecological bust. Caribbean refineries and the sharp uptick in supertanker traffic they invited to the region brought a new problem: coastal oil spills. From Florida to Guyana, the wider Caribbean experienced more than thirty major oil spills during the 1970s and countless mundane releases of petroleum from ships and shoreline facilities. The Gulf of Mexico and Caribbean region, it is worth noting, have hosted the four largest accidental oil spills in human history.[2] As pipelines leaked, wellheads blew out, refineries dumped effluent into lagoons, and supertankers discharged oily bilge or occasionally even collided with other tankers, all varieties of oil spill assailed coastal ecologies. By 1976, marine-bound crude oil was designated "the pollutant of highest priority concern to the Region" by the United Nations Environmental Program and a commission of Caribbean representatives (Atwood et al. 1987: 540). The resulting initiative, called CARIPOL, faced an unexpected difficulty in reining in hydrocarbon effluent: the Caribbean Sea was so "chronically contaminated" with petroleum that it was next to impossible to determine a natural baseline against which to measure and manage petro-pollution (545). The wider Caribbean rather abruptly found itself awash in spilled oil.

The latter half of this chapter describes the scientific response to the newfound problem of crude oil in the Caribbean Sea. As oil washed up in various Antillean locales, state-backed lawsuits brought sustained analytic attention to emerging concerns over the vulnerability and value of coastal ecologies. Tracking in and out of one prominent oil spill in Puerto Rico in 1973, this chapter argues that the scientific response to these coastal oil spills fundamentally reformed the meaning of mangroves. In conversation with recent scholarship that has taken up the "liminal" quality of mangroves as a sharply Caribbean analytic, I show how mangroves offer a telling window into the recent social history of the West Indies (Price and Price 1997; Ogden 2011). My approach differs slightly from such work, however, by emphasizing one of the material venues within which the relationality of mangroves first became factual and operable: Caribbean oil spills.[3] I show how the deleterious impacts of these spills became a kind of field laboratory for radically rethinking the agency of mangroves. The research done in the wake of such disasters grounded a new empirical appreciation for the ecological work and economic worth of mangroves in the Caribbean. This scientific valorization of mangroves undergirds much of the rising cultural celebration of mangroves as a new emblem of postcolonial identity in the Caribbean today.

This chapter, then, offers a Caribbean version of how nature continues to matter in the so-called Anthropocene. As the Anthropocene rattles and reframes scholarly debates over the constitution of modernity, fossil fuels have been taken up as the explosive bookends of industrial society. Fossil fuels first equipped industrial society to seize functional autonomy from the natural world (Sieferle [1982] 2001; McNeill 2000; Crosby 2006). As the planetary consequences of this divorce come into impending focus, fossil fuels are seen as a key geochemical driver in the vengeful rebounding of the natural world and the endpoints it heralds for modern life (McNeill and Engelke 2016). In this popular story of the Anthropocene, fossil fuels are the agent provocateur in the breathless genesis of modernity and in its cataclysmic terminus.

Yet the inaugurations and disruptions that have taken shape around fossil fuels are far from a singular beginning and end. While some suggest that the founding rupture of nature and society has rendered modernity uniquely incapable of acknowledging the planetary catastrophe of its own making (Latour 2004)—indeed, many scholars now locate the

FIGURE 12. Mangroves in St. Croix. Photo by author.

most pressing form of critique in an analytical alignment with Indigenous ontologies presumed to be outside the modern episteme (Kohn 2013; Viveiros de Castro 2014)—such arguments often overlook the complicated and contradictory terrain of environmental reflexivity within the modern project, much of it provoked by fossil fuels. Among other things, the negative ecologies of fossil fuels instigated fairly robust acknowledgment of the vitality of natural worlds within the modern project. The cresting consequences of fossil fuels have long contorted the basic biochemical conditions of life, whether in urban smog or acid rain, in hydrochlorinated pesticides, or now, through climate change. Tilting the conditions of life just beyond the fixtures of modern society, fossil fuels have opened the door to new understandings of and new obligations toward those newly precarious conditions. Fossil fuels have not done away with natural worlds; their destruction has unloosed new scientific and political desires for vital nature.

OIL REFINERIES AND THE (RE)MAKING OF THE MODERN CARIBBEAN

The studied Caribbean is, in many ways, a wager on the legacy of the sugar plantation. The late Sidney Mintz (1966: 925), resident dean of

Caribbean scholarship, wrote at the beginning of his career: "The Caribbean region has been both 'urbanized' and 'westernized' by its plantations, oil refineries, and aluminum mines, more than by its cities." Attention to oil refineries and aluminum plants has long slipped out of focus as the plantation came to be seen as the ascendant site in the making of the contemporary Caribbean. As told through the intersecting aims of ethnography, literature, and nationalism, this turn toward "plantation economies" qualified the conceit that the modern Caribbean was not the distant imprint of some colonial design but rather the negotiated outcome of a decidedly regional history (Lewis 1954; Best 1968; Beckford 1972; Best and Levitt 2009). The essential infrastructure of the plantation (coerced migration of foreign labor, rural concentrations of labor and capital, a conscripted modernity, and racial orderings of status) and its social consequences (a reconstituted peasantry, the unease between the state apparatus and national identity, and creolized modalities of identity) are widely seen as the preeminent examples of what makes the Caribbean a special and enduring region (James [1939] 1989; Williams 1944, 1970; Mintz 1966. 1975; Mintz and Price 1985; Trouillot 1992). And yet today, working plantations and export-oriented agriculture more generally are in noted decline across the region. The Caribbean Development Bank (2003: 4) recently reported that agriculture is in a "state of crisis" as the passing presence of tourists, finance, and petroleum have become the pillars of the region's economy (Sheller 2003; Mauer 2004; Hughes 2013).

As the physical presence of sugar plantations recedes, many scholars continue to insist on the "haunting continuities" that fix contemporary social life on the now-immaterial foundation of the plantation (Chatterjee, Das Gupta, and Rath 2010: 11). In today's Caribbean factories and data centers some anthropologists hear echoes of slavery, and they presume that the plight of the present is best understood by first overlaying it with the social forms of the plantation (e.g., Yelvington 1995; Freeman 2000). While the resulting insights can be fruitful, they often miss the generative manner in which the critique of the plantation itself has been used to justify alternative imperial interventions in the region. My point here is not to dismiss the plantation as a scholarly project and even less to deny the durability and mobility of the plantation's central architecture: racial hierarchies and single-use landscapes. Far from a simple rejection of the plantation, my aim, with reference to the sociology of critique (Boltanski and Chiapello 2007), is to describe how the critique of the plantation has, in itself, become an influential social actor in the contemporary Caribbean. Scholars in the Caribbean no longer have a monopoly on the critique of the plantation (if they ever did).[4] Over the past century, colonial governments and oil companies have argued that refineries could help overthrow the racial legacy of the plantation and catapult the Caribbean into a modern future.

During the twentieth century the Caribbean became a key energy outpost for imperial powers. This cardinal economic realignment remains underappreciated in scholarly and popular understandings of the region. As the Panama Canal brought new global shipping lanes to the region and as European navies and trading concerns retrofitted their fleets to run on bunker fuel, oil depots and refineries were built across the region (Ramsaran 1989). Unlike refineries built in the United States and Europe, designed to serve adjacent urban markets, these outsized Caribbean refineries were scaled to the oceanic merchant and military networks they supported. During World War II, the Royal Dutch Shell refinery on Curaçao became the largest refinery in the world, followed closely by Standard Oil of New Jersey's refinery on Aruba. These two massive Caribbean refineries provided over 80 percent of the Allies' aviation and naval fuel and attracted concerted attacks from German U-boats. Many of these early Caribbean refineries were designed to process Venezuelan and Mexican crude oil within "the solid European administrations" of Caribbean colonies (Hartog 1968: 308). As Fernando Coronil (1997: 107) noted, Venezuelan leaders actively encouraged the strategic placement of refineries in the Caribbean "in order to avoid creating large concentrations of workers with their attendant labor problems" in Venezuela.

In the postwar period oil companies in the United States faced a similar dilemma. An upsurge in worker strikes at domestic refineries joined with rising regulatory concerns over municipal pollution to encourage some firms to seek competitive advantage elsewhere.[5] For some oil companies, former plantation land and freedmen communities largely along the Mississippi River in Louisiana offered a racial exception to these new points of friction (Ottinger 2013; Misrach and Orff 2014). For other oil companies, the Caribbean became an attractive site to expand refining capacity while sidestepping the demands of organized labor and rising environmental oversight in the United States (Gorman 2001; Payne and Sutton 1984). The Caribbean's "political stability, its deep harbors, its lack of environmental regulations, and its proximity to major shipping lanes" provided an exceptional venue for offshored US hydrocarbon infrastructure (Barry, Wood, and Preusch 1984: 89). Such a move to the Caribbean also paralleled a fundamental reorientation of US energy infrastructure in the 1960s away from declining domestic reserves and toward a newfound dependence on imported crude oil.

Oil production in the United States peaked in 1970 (until fracking reversed the downward trend around 2010). A hemispheric event, peak oil in the United States also helped transform Caribbean refineries into an imperial circuit of the US energy grid. After World War II, abundant domestic reserves of oil underwrote a new American Dream of cheap food, big cars, and suburban ease. As the domestic flow of crude started sputtering in the 1960s and 1970s, the nation faced a dilemma: either recognize natural limits or obtain oil from elsewhere.

In the 1970s, the United States debated whether to redesign American life around alternative sources of energy, efficiencies achieved through public investments in building design and transportation, and drastically curtailed military expenditures of fuel (all of which were key platforms of the first Earth Day in 1970). Or, in the other direction, to throw the weight of the federal government into a more imperial pursuit of foreign oil. President Nixon, opting for the latter, helped deepen the American addiction to fossil fuels far beyond what the country itself could provide. In 1955, roughly 90 percent of petroleum consumed in the United States came from domestic sources. By 1977, roughly half of the gasoline, jet fuel, and heating oil consumed in the United States came from foreign oil. This rising American dependence on foreign oil transformed the Caribbean into the premier refining hub of the eastern United States.

Until Nixon lifted it in 1973, domestic refineries were bound by the Mandatory Oil Import Program, which imposed strict quotas for imported oil. (Designed to minimize dependence on foreign oil, the program set a maximum level of imports at about 12 percent of domestic demand.) In 1965, US territories in the Caribbean were granted exemptions from these quotas, and soon refineries in Puerto Rico and the US Virgin Islands became an advantageous route for cheaper imports to slip into the United States. Moreover, that same year the US Congress authorized a series of tax exemptions that encouraged domestic oil companies to build new export-oriented refineries and petrochemical plants in Caribbean territories (Dietz 1986). Over the next two decades, US companies built more than a dozen entrepôt refineries on Caribbean islands.[6]

While the exceptionality of colonial territories provided their opening advantage, other events shored up the importance of Caribbean refineries. The World Bank and other international organizations actively encouraged Caribbean states to welcome this new "enclave-type" processing of petroleum products for export to the United States as a crucial step in developing Caribbean economies and disciplining its societies into the expectations of the modern economy (a vision the World Bank also applied to Singapore) (Chernick 1978: 139; World Bank 1984).[7] The OPEC embargoes against the United States in 1973 and 1979 further consolidated the strategic importance of Caribbean refineries as they acted as a back door for Middle Eastern oil to "leak" into the United States (Middle East Research and Information Project 1974: 23). By 1990, roughly one-sixth of the oil consumed in the United States and "over half of the refined petroleum imported to the U.S.—including oil from African and Middle Eastern sources" passed through Caribbean refineries (Richardson 1992: 116).

To summarize: between 1950 and 1990 Caribbean oil refineries re-made the political economy of the region and constructed an exceptional pathway for imported crude oil and petroleum products to enter the United States. These Caribbean refineries, then, played a pivotal, if largely unrecognized, part in the imperial realignment around the properties of fossil fuels so aptly described by Timothy Mitchell in *Carbon Democracy* (2011). Around crude oil, Mitchell argues, the constituent field of empire changed from the racial ordering of labor to the techno-political ordering of energy flows (207–8). Joining domestic desires for energy-intensive lifestyles with a realignment of global energy infrastructures, crude oil heralded a fundamental shift in the texture and technique of US empire. These "new and less visible forms of imperialism," as C. Wright Mills described the changing scene (1959: 4), brought renewed importance to the Caribbean. The US territories in the Caribbean became primary sites for retrofitting the US empire around the oceanic distribution of crude oil. These island territories became, as one recent appraisal put it, "critical nodes" in the "networked empire" of contemporary US power (Oldenziel 2011: 13).

While a number of scholars have examined the growing imperial inflections of fossil fuels in this era, they have often done so by showing how oil companies worked to violently safeguard foreign extraction sites from local discontent and nationalized outcomes (Watts 2005; Mitchell 2011). Yet the infrastructure that exempted crude oil from democratic concerns was far more expansive than policed wellheads and buried pipelines. Supertankers provided an unprecedented degree of flexibility. Refineries and petrochemical plants built in communities of color on freedmen townships in Louisiana and in coastal ports in Caribbean territories constructed an exception to the growing rights of workers and the environment (Ottinger 2013; Misrach and Orff 2014).[8] Militarized drilling sites, supersized tankers, enclave refineries, and suburban lifestyles worked in concert in the "imperial formation" that took shape around crude oil during the twentieth century (Stoler and McGranahan 2007). Scholarship that presumes that the imperial imprint of oil unfolds only within the geography of extraction can lose sight of the extensive investments in distribution, refining, and consumption that make the empire of oil possible.

As US oil companies found the colonial status of many Caribbean ports advantageous to their global operations, the critique of the previous modality of empire—crystallized in the image of the sugar plantation—provided salient local justification for aligning Caribbean islands with an emergent modality of empire: the enclave refinery. In Puerto Rico and the US Virgin Islands, the overthrow of the plantation was the leading argument made by the colonial government to welcome and legitimate the arrival of refineries and petrochemical plants. In St. Croix, for example, Hess Oil built the world's largest petrochemical plant and second largest oil refinery in 1966. Such heavy industry would usher in, in the words of the appointed governor, a "bloodless revolution" that deposed the racial feudalism of the plantation and ushered in the color-blind modernity of industrial capitalism (Thurland 1979: 167). The premise was flawed, and the promise failed. In the early 1960s, merchants and unions began a sustained campaign to finally overcome what the head of the Virgin Islands Labor Union called the "economic slavery" of agriculture (*Daily News* 1962c: 10). The urban merchant class (with the help of the national Democratic Party in the United States) decided that aluminum and petroleum plants were the islands' future, or at least the future of their own interests, since many local farmers had become quite adept at bypassing their levied mediation. Many elderly farmers I spoke with in St. Croix recalled this moment with delight, relishing their independence from an urban elite. One retired merchant I interviewed had a different take: with noted displeasure, he criticized the insular attitude of these farmers, who lived and worked without regard for exports.

To right this, from 1961 to 1967 colonial governor Ralph Paiewonsky made executive agreements, first with Harvey Aluminum and then with Hess Oil, to bring heavy industry to the southern shore of St. Croix.[9] To encourage such development, the colonial government donated several hundred acres along the coast to the two companies and excused both from paying local taxes and following existing energy importation rules. They also expropriated land from local farmers and collective farms. In his memoirs, Paiewonsky reminisced that his primary goal in bringing industry to St. Croix was to transform the Virgin Islands "into a modernized Western society. [. . .] As a businessman myself, it was clear that my sympathies would be on the side of business" (Paiewonsky and Dookhan 1990: 219–20).

Under the banner of overcoming the regime of the plantation— "Governor Plans to Wipe Out St. Croix Feudal System" ran one headline (*Home Journal* 1964: 1)—the colonial government actively contrasted the racial history of the plantations to the modern promise of the aluminum and petroleum industries. Such a contradistinction neatly overlooked the vibrant present of many island farmers.[10] In a particularly nasty turn of events, colonial authorities seized the most fertile swathe of land on the island from small farmers and handed it over to the industrial coalition, claiming "St. Croix has had a sugar economy for long enough."[11] Most of the farmland usurped was simply fenced off and left empty for the next fifty years. Occasionally Hess Oil, which soon controlled the land, would sell a section back to the local government to build a prison or a public housing complex. One large section was paved over to welcome the island's first shopping mall. As farmers and their allies protested throughout the 1960s, the colonial government routinely asserted that only the wages of modern industry could emancipate the island from the racial scourge of the plantation (*St. Croix Avis* 1963; *Home Journal* 1964). It did not work out that way.

After the first batch of industrial workers recruited from former farmers on St. Croix went on strike demanding the wages and benefits they had been promised, both Harvey Aluminum and Hess Oil fired most of their native workers (*St. Croix Avis* 1964; *Daily News* 1965a). One retired refinery official offered an airbrushed version of this history, stating, "Crucian blacks were unemployable at the refinery because they preferred to work for the government."[12] The companies began importing Afro-Caribbean workers from other Caribbean islands as a temporary workforce that could be politically disenfranchised and easily deported.[13] While unemployment among the island's native residents was 3 percent in 1960, it spiked to over 10 percent in the years after the refinery and petrochemical plant arrived (Miller 1979). To stave off this crisis, the government began hiring, and it soon surpassed the refinery as the island's largest employer (Miller 1979). Flush with oil tariffs— by 1977 the Virgin Islands was processing roughly $2.5 billion worth of petroleum products, while all nonpetroleum exports totaled only $70 million—St. Croix's petro-infused government became the largest employer on the island (Miller 1979). "The public sector comprised the largest portion of total employment," one study reported (Tri-Island Economic Development Council 1983: 22), noting that almost 40 percent of jobs in the Virgin Islands were in government. One employment study even suggested that during the 1970s over 75 percent of the new jobs created in St. Croix for citizens were in the public sector (Pobicki 1980). While the oil industry did little to help ordinary people (and much to harm them), it did create a sprawling government bureaucracy and, at least on paper, turned St. Croix into a robust economy. From the 1970s well into the 2000s, the US Virgin Islands—based solely on what passed through this single megarefinery—was regularly listed as one of the top ten sources of oil imported by the United States.

While the story of how St. Croix aligned with the new imperial geography of US energy flows is particularly egregious, it is not an isolated incident. After the US Congress created special tax exemptions for refineries and petrochemical plants built in Puerto Rico in 1965, oil companies like Tesoro, Sun Oil, Gulf Oil, Union Carbide, and Philips Petroleum constructed new facilities on the island's southern coast, largely designed to process Venezuelan oil and then ship the refined petroleum products to major cities on the Gulf Coast and Eastern Seaboard (Dietz 1986). That same year, the government of Puerto Rico declared petroleum refining and petrochemical industries to be the island's "top industrial priority" (quoted in Whalen 2001: 32). US oil companies also built new enclave refineries and transshipment centers in Aruba, Antigua, St. Lucia, and the Bahamas (Paget 1985; Ramsaran 1989). US firms increased their investments in the petroleum sector in the Caribbean by 400 percent during this period, and by 1980 the petroleum industry was "the largest U.S. direct investment in the Caribbean" (Barry, Wood, and Preusch 1984: 19). Across the Caribbean and on islands with no crude oil reserves of their own, "the fast growing refinery and petrochemical industry" promised to become, as one report on Caribbean development put it, "a focal point of the island's further industrial development" (Powell 1973: 39). As in St. Croix, this goal proved elusive. On many Caribbean islands the building of such refineries led to spiking unemployment, a metastasized state bureaucracy, and social unrest (Paget 1985; Pantojas-Garcia 1990).[14] While less commented upon, the introduction of enclave refineries also dramatically remade the coastal landscape of the Caribbean.

As colonial and national governments across the Caribbean aligned their futures with the promise of expanding hydrocarbon networks, "mangrove swamps" were widely seen as an opportune place to build. Brazil, Mexico, Venezuela, and many Caribbean nations embarked on mangrove eradication programs during the 1950s and 1960s, in many cases with the express purpose of developing industrial ports (Mumme, Bath, and Assetto 1988; Miller 2007). Mangroves, asserted a 1967 report on development in the US Virgin Islands, were "cesspools of disease" that should be destroyed to make room for more productive applications like oil refineries and petrochemical plants (Virgin Islands Office 1967: 3). Alongside tax breaks and suspended regulations, the infamous "industry by invitation" (Lewis 1950) in Puerto Rico also rested on a more literal foundation: bulldozed and backfilled mangroves.

In St. Croix, the colonial government viewed the mangroves as it viewed the plantation: as a lingering anachronism best cleared out so the future of industrial modernity could finally arrive. Torn from the present, repressed agriculture and razed mangroves provided the physical coordinates that proved progress was happening. They both became history. When he announced the arrival of a world-class refinery on St. Croix, Governor Paiewonsky proudly noted that construction would wipe out the island's largest mangrove forest. The mangroves were "worthless," he said, noting that the area was "infested with mosquitoes and sand flies. It cannot by cultivated. But with this plant there, property values over the area will be enhanced" (*Daily News* 1962a: 2). "Where a wild mangrove swamp once defeated practical land use, a new deep water port has been dredged," one booklet about St. Croix development proclaimed, as it invited other industries to the island (Virgin Islands Office 1967: 32). The government paid Harvey Aluminum $500,000 annually for the cost of dredging the port and clearing the area of mangroves (*Daily News* 1962b: 1), an arrangement later expanded and extended to Hess Oil. A major tanker oil spill in 1971, along with chronic leaking at the bauxite facility and oil refinery, further assailed the once vibrant tidal forests of St. Croix (*Daily News* 1971: 2). At the time, such destruction of the mangroves was tolerated if not celebrated as evidence of progress.

OIL SPILLS AND THE CHANGING MEANING OF MANGROVES

In March 1973, the oil tanker SS *Zoe Colocotroni* left Venezuela full of crude oil for a refinery in Puerto Rico. With its navigation system broken, the ship proceeded by celestial reckoning. Eight hours after clouds obscured its navigators' view of the stars, the tanker slammed into a reef on Puerto Rico's southwestern coast. The captain, after repeated attempts to reverse the tanker off the reef, ordered the crew to lighten the ship by dumping its load of crude into the sea. Soon the ship was dislodged and on its way. The next morning a large oil slick moved into the Bahía Sucia estuary, drifting into an extensive mangrove forest. Although much of the oil was eventually removed, many mangroves withered and began to die.

Local leaders and newspapers initially expressed relief: it was only mangroves that had been impacted. The oil slick had not reached any towns or popular beaches but instead had come ashore on a relatively uninhabited section of the coast. As marine biologists and a new environmental agency examined the site of the spill, however, they began to question that assumption. In the sudden absence of mangroves, disruptions to coastal life rippled outward. Puerto Rico commissioned a number of new studies to determine not only how the oil killed the mangroves but also the role, more generally, of mangroves in marine life. These studies and others like them laid the groundwork for a new empirical appreciation of the integral relationality of mangroves and their role in sustaining coastal ecologies and economies. Two years later, they also prompted Puerto Rico to file suit against the owners of the SS *Zoe Colocotroni* not only for the cost of cleaning up the oil but also, advancing a new kind of legal accusation, for the destruction of mangroves and the ensemble of marine organisms they fostered.

The matter of guilt was established early on. The ship was, after all, operating with the wrong maps, a damaged navigation system, and, as the court put it, "an incompetent crew."[15] The legal debate that unfolded hinged on a separate issue: How can, or how should, the courts value mangroves? This question was key to the lawsuit and, at the same time, largely without precedent, for the bigger question was quite explicitly: How much is nature worth? While economists have long theorized about "natural capital" (e.g., Hotelling 1931), and common law and civil law both have a robust tradition of assessing damage to public resources like forests or fisheries, this case moved in a different direction. Instead of valuing nature through what *homo economicus* might make of it, this case asked how modern society infringed upon the independent life support systems of nature. That is, it inquired into nature not as a potential commodity lying in wait but as a vital productivity in its own right. "To say that the law on this question is unsettled is to vastly understate the situation," one judge remarked.[16]

Mangroves, as we now know, form one of the world's most robust ecosystems. Characterized by their "strict fidelity" to the tidal zone in the tropics, they thrive in estuaries and other areas where fresh water mixes freely with the ocean (Tomilson 1986: 3). This ability, in turn, transforms that churning line separating land and sea into vibrant habitat. Mangroves establish "interface ecosystems, coupling upland terrestrial and coastal estuarine ecosystems," drawing each into an unparalleled flourishing (Lugo and Snedaker 1974: 60). They are an emblematic case of what biologists now call the "emergent properties" of complex systems: that is, "those properties that arise from a system's components acting in concert and may not be readily identified or understood by the study of those components in isolation" (Feller et al. 2010: 397). By bringing solar energy, atmospheric elements, dissolved nutrients, and a spectacular array of organisms into concert, mangroves foster an exponential increase in the productivity of the ecosystem. The resulting ecological society far exceeds the sum of its parts.

Recent UNESCO studies have found that, worldwide, "80 percent of marine catches are directly or indirectly dependent on mangroves" (Kjerfve, Lacerda, and Diop 1997: vi; see also Ellison and Farnsworth 1996). Although mangroves comprise less than a single percent of the earth's surface, one recent estimate suggests their ecological import touches roughly half of the planet's natural resources (Costanza et al. 1997). An astounding variety of birds, fish, shrimp, crabs, and other marine life thrive in the brackish tangle of mangroves. As Odum, McIvor, and Smith wrote (1982: 86), "At no cost to man, mangrove forests provide habitat for valuable birds, mammals, amphibians, reptiles, fishes, and invertebrates and protect endangered species, at least partially support extensive coastal food webs, provide shoreline stability and storm protection, and generate aesthetically pleasing experiences." Teeming with life, mangroves have become the premier example of ecosystem services in a growing field of scholarship devoted to accounting for the worth of nature's agency in our present (Costanza et al. 1997). The more we learn about mangroves, the more we find ourselves in awe of their local and planetary significance.

Where did this rising appreciation of mangroves come from? It was, after all, not that long ago that mangroves were almost uniformly seen as a drag on development, and a scientific outlier article in *Nature* described mangroves as the "freaks" of the natural world (Davis 1938: 556); until quite recently many marine scientists found terms like "swamps," "wastelands," "curiosities," or even "depauperate" appropriate for describing mangrove forests. John Steinbeck, on a marine science expedition, summarized the sentiment: "No one likes mangroves" (1951: 101). Today, a very different understanding has taken hold. Mangroves are widely celebrated as one of the earth's most remarkable features. While many appreciations of them suggest that our rising regard for tidal forests is part of the progressive enlightenment of science, such a narrative misses the uneven material grounding of this new ecology of mangroves. Far from the steady march of reason, much of the research that gave empirical momentum to the vitality of mangroves took place in the wake of their crude eradication in the Caribbean.[17]

Mangroves once flourished along the coasts of every Caribbean island and populated the Atlantic Coast from New Orleans to Buenos Aires (Lacerda et al. 1993; Dean 1997; Miller 2003). Today, they are in spectacular decline. Recent estimates suggest that well over 50 percent of the world's mangroves have been destroyed, the bulk since 1950 (Feller et al. 2010). Today, the total extinction of mangroves remains a plausible future event (Duke et al. 2007). In the wider Caribbean this collapse is particularly stark (Ellison and Farnsworth 1996). Panama lost 40 percent of its mangroves by 1980 and Puerto Rico 85 percent, while in Venezuela some estimate that only 10 percent of the original mangrove forest remains (Lugo and Cintrón 1975; Ellison and Farnsworth 1996). As a botany textbook explains: "The chief factor that currently modifies mangrove distribution is the activity of industrial man" (Tomilson 1986: 61). While surprisingly resilient to changing conditions in the water and temperature and even to a variety of human encroachments, mangroves are extremely susceptible to the main ingredient of contemporary capitalism: crude oil. This "Achilles heel" of mangroves, as one report described it (Odum and Johannes 1975: 54), offers a novel interpretation of the crisis of tidal forests in the Caribbean: the rise of oil refining in the Caribbean mirrors the collapse of mangroves.

While many Caribbean refineries were built atop reclaimed tidal forests—a 1982 conference on protecting the tidal forests around the Hess refinery on St. Croix noted that until very recently "mangrove management meant reclamation" (Cintrón and Schaeffer-Novelli 1982)—the physical construction of refineries was only the beginning. As new refineries brought an influx of tanker traffic to the region, routine ballast discharges and an escalating series of tanker accidents released crude oil into the coastal environment. Refineries were also notoriously leaky; one study estimated a midsized refinery in 1960 leaked about fifteen thousand barrels of oil a year (cited in Gorman 2001). During the 1970s, the wider Caribbean experienced more than thirty major oil spills and countless ordinary leaks and discharges. In 1979, two of the largest oil spills in human history unfolded as back-to-back disasters in the region. In February, two fully loaded supertankers collided in heavy fog just off the coast of Tobago, and as explosions ripped the ships apart, they spilled nearly 2.1 million barrels of crude oil. Less than a month later, an exploratory well in the shallows off the coast of the Yucatan experienced a blowout. Over the next 385 days, the Ixtoc I wellhead spewed approximately 3.6 million barrels of crude oil into the Gulf of Mexico. While the 1970s was capped with spectacular disasters, a litany of more modest spills and more mundane accidents continued to mar the marine environment of the Caribbean.

These oil spills and the damage they inflicted on coastal ecologies provided a telling environmental register of the realignment of Caribbean islands around imperial petro-networks. As the United States recalibrated its energy flows around imports and the spatial exceptionality of island colonies in the Caribbean, supertankers and offshore refineries became primary instruments in its petro-networks. While the pipeline may have opened the door to a new logic of energy infrastructure that bypassed labor and democratic action, as Timothy Mitchell (2011) has argued, supertankers and enclave refineries extended that flexibility into a new imperial apparatus. They also, however, introduced a new problem that came to attract growing public attention: coastal oil spills. "Petroleum has become a devil in our civilization," warned a 1967 *New York Times* feature on marine oil spills: "Whether in a single dramatic incident or slowly, by default, it is fouling the seas, creating a survival issue both for sea life and for man himself" (Rienow and Rienow 1967: 25). By the early 1970s, and to the noted surprise of many observers, it was found that "the vast majority of United States oil spill incidents occur within coastal waters" (Gundlach, Hayes, and Getter 1979: 90). Petroleum in the marine environment was fast becoming a domain of official concern, as documented by a series of conferences and reports commissioned by the National Research Council (1975, 1985). This new oceanic orientation also registered in popular culture as the dominant imagery of oil spills in magazines and newspapers began shifting from gushing wellheads to sea birds coated in crude (Morse 2012). These oil spills invited unexpected scientific documentation and expansions of state responsibility to the emerging maritime routes of US energy networks. The marine environment was, in many ways, coming to replace organized labor as the premier point of friction in these imperial petro-networks.

In the Caribbean, these concerns came into sharp focus around the impact of oil on mangroves. Scientists who studied how oil spills affected Caribbean mangroves found descriptions like "catastrophic" and "devastating" entirely appropriate (Burns, Garrity, and Levings 1993; Dodge et al. 1995). "There is no question," a leading ecologist summarized, "oil kills mangroves" (Lugo 1980: 51). One study found 96 percent mortality among juvenile mangroves following diluted exposure to petroleum pollution, compared with near zero mortality in unoiled sites (Grant, Clarke, and Allaway 1993). In wide-ranging studies of how various coastal ecosystems are affected by spilled oil—many of which were conducted in the aftermath of Caribbean incidents—tidal forests quickly achieved the title of most vulnerable (Rützler and Sterrer 1970; Odum and Johannes 1975; Gundlach and Hayes 1978).[18] Due to their sheltering nature, "mangrove forests are routinely sites where oil accumulates after a spill" (Lewis 1983: 171). Once accumulated among the buttressing roots, crude oil severely impairs the respiration of mangroves through a sort of induced "mechanical suffocation" (Snedaker, Biber, and Aravjo 1997: 2). Coating the roots and rhizomes of mangroves, crude oil effectively strangles tidal forests from the nutrient exchanges and biogeochemical cycles they depend upon, leading to "severe metabolic alterations" (Odum, McIvor, and Smith 1982: 80). The impact can be quite sudden. Oiled mangroves often defoliate and die within a matter of days.

Many of the first ecological surveys of these oil spills focused on a "dripping oil and dead-body count approach," as one retrospective review put it (Snedaker, Biber, and Aravjo 1997: 1). Yet in the sudden and often persistent absence of mangroves after a spill, a more expansive definition of environmental harm took shape. For one thing, the impact was surprisingly long-lasting. For example, in 1968 a tanker broke up in a storm off Panama's coast, and slicks of crude wiped out whole sections of mangroves (Rützler and Sterrer 1970; Birkeland, Reimer, and Young 1976). Nearly thirty years later the estuary still bore the imprint of the injury: while the mangroves had begun recolonizing the estuary, trees in the heavily oiled areas remained visibly shorter and "with less overall biomass" (Duke, Pinzon, and Prada 1997: 9). That same study concluded that the durable impacts of oil spills on mangrove forests often cover an area five to six times larger than the area of immediate lethality (9). As tidal forests failed to heal after spills in the Virgin Islands and Puerto Rico, scientists documented how mangrove skeletons and sediment foster anaerobic processes that concentrate the toxicity of crude oil in the marine environment for years, if not decades (Lewis 1983; Corredor, Morell, and Castillo 1990). In 1986, a refinery accidentally released crude oil into a mangrove-lined lagoon on the Caribbean coast of Panama. A decade later, biologists described a wound still festering: the enduring impact was so apparent "that the affected site exhibited the appearance of having been subjected to an explosion" (Snedaker, Biber, and Arajo 1997: 3).

Moreover, the impact extended beyond the physical forest. After an oil spill, the sudden absence of mangroves threw the quiet services they provided into stark relief. In 1976, Columbia witnessed local fisheries enter into a sustained period of decline after a tanker spill wiped out coastal mangroves (Hayes 1977). Other fisheries adjacent to oil-impacted mangroves in the Caribbean reported similar declines or collapses of species "independently of any effects of hydrocarbons on the organisms themselves" (Garrity, Levings, and Burns 1994: 327). In other areas, the years following an oil spill witnessed extensive coastal erosion after stricken mangroves loosened their grip on the shoreline (Dodge et al. 1995). These oil spills offered an effective window for witnessing how mangroves contributed to coastal ecologies and economies, from providing a crucial habitat for juvenile shrimp, crabs, fish, and other commercial species, to filtering agricultural or urban runoff, to absorbing dangerous storm surges.[19]

The oil spills that beset the Caribbean triggered pioneering scientific studies of mangroves as such. In a curious way, oil spills grounded a new ecological appreciation of mangroves. Mangroves do not fit neatly into the given taxonomies of species or commodities, but the ways in which they do not fit turn out to be of crucial importance for the ensemble of life gathered within tidal forests. "Unlike other terrestrial communities that can be lived in, managed, or exploited by man, mangroves offer only a few direct uses, which may account for man's historical ambivalence concerning their value," observed one of the first major scientific review articles on mangroves (Lugo and Snedaker 1974: 39). This article was written by two scientists who had studied the destruction of mangroves in the *Zoe Colocotroni* spill in Puerto Rico and provided expert testimony in the subsequent trial. In fact, many of the principal early ecological studies of mangroves were instigated (and funded) not by strictly academic concerns but by new legal questions regarding injury inflicted by Caribbean oil spills (e.g., Odum and Johannes 1975; Nadeau and Bergquist 1977; Gilfilian et al. 1981; Lewis 1983; see also Odum 1970 for a more academic arrival at the ecology of mangroves). Taking a wide range of Caribbean oil spills as a coherent field of study (Getter, Scott, and Michel 1981), these disasters were involved in voicing and valorizing a new understanding of the agency of the natural world in an era of hydrocarbon endangerment.

This turn to oiled mangroves is but one regional constellation of a much wider shift in the natural sciences. In the postwar period, the natural sciences began to study life not in some contrived isolation from modernity, but as altered by modernity (Beck 1986). While much has been made of the rise of biotechnology in facilitating this shift (Rabinow 1996; Rose 2006), perhaps the more substantial historical subject has been the material afterlives of nuclear weapons and hydrocarbon fuels. We are just beginning to grasp the key role radioactive fallout played in enabling and equipping the earth sciences (Masco 2010). The parallel role of hydrocarbon pollution in shaping the object and practice of environmental science—indeed in providing an empirical outline of "the environment" itself (Bond 2013)—has yet to receive the critical and comparative attention it deserves. So much of what we know of, and how we have come to care for, the conditions of life like clean air and clean water and now a stable climate, rests on how fossil fuels first disrupted them. These insights are far from a universal process of enlightenment. The analytic and ethical definitions of vital nature instigated by fossil fuels remain uneven and rooted in particular experiences of harm. In the Caribbean, this had everything to do with oil spills and mangroves.

The *Zoe Colocotroni* case was argued in the courts for over a decade and came to catalog the changing meaning of mangroves. As one of the defendants complained, these new questions had made "a court case, not out of the oil spill itself, but of the biological effect of the oil spill" (quoted in Lugo 1980: 55). The legal debate in this case came to rest on the value of mangroves. The market, the court concluded early on, was not the best means to assess the value of nature, stating: "Many unspoiled natural areas of considerable ecological value have little or no commercial or market value."[20] From fresh air to clean water, the court recognized that certain vital elements of life exist outside of market valuation. As petroleum infringed upon these independent life support systems, the court was pressed to come up with a convincing method to calculate damage to them. Instead of real estate, the court turned to science. Marine biologists, questioned again and again on how mangroves matter, answered in ways that shifted the measure of value from the indexed exchange to ecological work. When asked by trial lawyers how much mangroves were worth, marine biologists described what mangroves do. Walking the court through the new studies of mangroves—a great many of them conducted in the wake of Caribbean oil spills—these marine scientists testified to the centrality of mangroves in coastal ecologies. "The mangrove components of these systems are of prime importance," the court eventually concluded. "These areas are breeding, feeding, and nursery grounds for substantial populations. [. . .] Additionally the mangroves themselves, and in particular the red mangrove, are the primary food-producing agents of the organic materials available to the aquatic food chain."[21]

The final *Zoe Colocotroni* verdict was written in 1979, the same year that the two record-breaking oil spills hit the wider Caribbean. "In recent times," the ruling stated, "mankind has become increasingly aware that the planet's resources are finite and that portions of the land and sea which at first glance seem useless, like salt marshes, barrier reefs, and other coastal areas, contribute in subtle but critical ways to an environment capable of supporting both human life and the other forms of life on which we all depend."[22] An earlier guilty verdict was upheld, and the court awarded Puerto Rico a record $6 million in damages to restore twenty acres of mangroves.

As oil development and oil disasters impaired the coastal ecologies of the Caribbean, a new definition of mangroves emerged. Their exuberant productivity—long known to marginal coastal communities (Miller 2003)—gained new scientific and political intelligibility through the encroachments of crude oil. Today, many nations and territories expressly protect mangroves. Thanks to recently changed laws and expansions of government authority in the Bahamas, Guyana, Panama, Puerto Rico, Trinidad, and beyond, countries are now acting to safeguard their tidal forests. In doing so, they often first reference the ecological services that mangroves provide to coastal communities (Lugo 2002). As climate change brings new attention to coastal vulnerabilities, many Caribbean nations are working to align their coastal infrastructures with the labor of mangroves (Vaughn 2017). On many islands, the image of the mangrove has become synonymous with Caribbean environmentalism. The US Virgin Islands recently began giving an annual award for the most environmentally friendly organization operating there, and the trophy is in the figure of a mangrove.

The mangrove, of course, has a longer presence in the Caribbean imaginary. "In brackish dialect" and "boiling with life," mangrove forests form the unfolding easel at the center of Derek Walcott's (1986) incisive 1973 poem, "Another Life." "*Mangrove reste un miroir,*" wrote Aimé Césaire (1990) in his 1982 poem "La condition-mangrove." "*La dodine celle du balancement des marées*" (The mangrove is a mirror. . . .

The rocking chair at the balancing of the tides). But as Richard Price and Sally Price (1997) have observed, the mangrove has begun shifting from poetic backdrop to insurgent symbol in the postcolonial Caribbean. Today, the figure of the mangrove is wielded across the Caribbean to mobilize the tangled histories of the region to reimagine political belonging in the present. As Caribbean writers turn to the mangrove as an emblem of postcolonial identity, they often do so by first consulting new ecological recognitions of the mangrove.

In Édouard Glissant's [1981] (1999: 67) elegant reflections, postcolonial Caribbean identity is folded into the ecology of mangroves: "Submarine roots: that is free floating, not fixed in one position in some primordial spot but extending in all directions in our world through its network of branches". In their trenchant treatise Éloge de la Créolité, Jean Bernabé, Patrick Chamoiseau, and Raphaël Confiant describe the ascendant qualities of being Creole as "la mangrove profonde," the profound mangrove (1989: 51). They write: "La Créolité est notre soupe primitive et notre prolongement, notre chaos original et notre mangrove de virtualités" (Creoleness is our primordial soup and continued sustenance, our founding chaos and our mangrove of possibilities) (28) This image of the mangrove has been artfully mobilized in the créolité literary movement across the Caribbean Basin. In such poetics and politics, the mangrove has come to evoke an anti-essentialist modality of life in the Antilles. "This land is mangrove," writes Raphaël Confiant. "The people are mangrove. The language is mangrove" (quoted in Price and Price 1997: 24).

This cultural mobilization of the ecological mangrove has also been put to work in social research. Richard Price and Sally Price (1997) have found the mangrove to be a uniquely Caribbean rhizome from which to theorize the present, as have others. "The metaphor of the mangrove guides my theoretical argument," writes Laura Ogden (2011: 90) in her investigation of the "confusing, nonlinear networks" (30) in the entanglements of the Florida Everglades. On one level, the mangrove bears resemblance to Eduardo Kohn's Amazon forest, an "emergent and expanding multilayered cacophonous web of mutually constitutive, living, and growing thoughts" (2013: 77). Like the Amazon, tidal forests are seen as a vital life force still outside the modular purview of modern purpose. As such, these unbowed forests are uniquely suited to voicing critiques of modernist orderings of people and landscapes. Verdant and vibrant places like the Amazon rainforest and Caribbean mangroves offer proof that another ontology is possible, one in which cooperation trumps competition, biosemiotics defy the sovereignty of the liberal individual, and the emergent qualities of interaction surpass any inscribed hierarchy of species. On another level, however, these properties of the forest are brought into political being by the very forces they stand against. Regenerating in the tides of empire, the ontological purchase of the Caribbean mangrove grows out of the disasters of oil imperialism.

The empirical opposition many ontological arguments presume between the disenchanted modern and its spirited opposite are often effective to the extent that they avoid historical questions of encounter (or more often, bracket the historicity of difference as a uniquely modern curiosity). In this, ontological venerations of alternative ecologies can sidestep histories of empire, which so often provide the grounds upon which epochal oppositions of modern and its other were first articulated and violently inscribed on people and landscapes (Wolf 1982). As Ann Laura Stoler and Tim Mitchell have demonstrated time and again, the modern project has always unfolded within a wider colonial field. Theoretically robust oppositions of the modern and its other can easily lose sight of this wider field, taking the effects of colonial encounters as the starting point of scholarly critique without inquiring much into their formation (Bessire and Bond 2014). The ecology of the mangrove may very well offer a scathing evaluation of modernization in the Caribbean, but we should be skeptical of claims that authorize their critiques by the purity they claim from the histories of our present. The ecological mangrove did not precede the empire of oil in the Caribbean. The ecological mangrove became a forceful counterpoint within and against the destruction of that imperial project.

As the mangrove becomes globally celebrated in this new era of planetary precarity, perhaps reassessing the imperial histories that enliven the ecological mangrove might once more bring the Caribbean to the fore in the making of the contemporary. The Caribbean has long been recognized as a historic crucible in the formation of colonial modernity and its creolized discontents (James [1939] 1989; Mintz 1986). With reference to refineries and mangroves, this chapter has argued that the Caribbean is also an unfolding crucible in the formation of petromodernity and its ecological discontents.

THE ENDS OF OIL

This chapter has linked up the imperial history of fossil fuels in the Caribbean with the disastrous history of the ecological mangrove as one regional articulation of a wider dialectic of hydrocarbon risk and environmental responsibility. Describing the local histories of aligning the Caribbean with US petro-networks and the ecological fallout of that alignment, this chapter argues that fossil fuels have done much to reorganize the region's economic and environmental landscapes. These concerns continue to work themselves out in consequential ways. Over the last few years growing environmental actions—many centered on the region's imperiled mangroves—have helped close oil refineries in the Caribbean. Between 1992 and 2009, the five major refineries on Puerto Rico ceased operation. "I helped shut down the last refinery on Puerto Rico," one EPA official proudly told me. Recalling a litany of environmental problems, he added: "These refineries were built on sensitive coastlines that should never have been developed. They should have been set aside and protected." Over the past few years, Caribbean islands that once linked their future prosperity with the imperial energy networks of the United States have found themselves further isolated from that receding future. Yet as environmental concerns have unexpectedly circumscribed the imperial networks of crude oil, other possibilities are taking shape. In the past decade, many renewable energy organizations have been looking to the Caribbean as a new laboratory for green energy. Due to the astronomical cost of petro-electricity, many islands are among the few places where renewable energy can compete with fossil fuel energy without subsidies.[23]

A CLIMATE CRUCIBLE

Today, St. Croix stands in a climate crucible. In 2012, HOVENSA, the mega-refinery that once shipped more oil to the United States than did Kuwait, was abruptly shut down.[24] This massive oil refinery underwrote economic development on this US territory in the Caribbean for fifty years. Today, the aftershocks of fossil fuels—whether in a legacy of toxic pollution or the rising fury of superstorms—threaten life on St. Croix. Facing up to these challenges is a costly affair, and many worry over how to build a future that can rise above the throes of petro-capitalism. Backed by pedigreed expertise, one plan gains momentum: heavily subsidize the rebooting of the refinery in the hope of bending the fiscal properties of fossil fuels into new investments in climate resiliency. For many civic groups and residents, such a plan seems the very definition of foolhardy. Battered and bruised by the environmental properties of fossil fuels, citizens demand a forceful break with the refinery to secure a more sustainable future today. The situation in St. Croix raises a crucial conundrum many of us are beginning to inhabit: Just how does oil end?

During its heyday, HOVENSA generated enough revenue to transform the modest island of St. Croix into a paragon petro-state. On paper, the economy flourished as the territorial government was flush with refinery tariffs, almost magically able to hire its way through every economic downturn. As is so often the case, such fiscal wealth came at tremendous ecological cost. The explosions of 2011 helped bring HOVENSA's imperial place on the island—its exported profits and gathered injuries—into unsettling focus. As explosions continued, EPA investigators were sent in. They soon uncovered long-standing practices of deferred maintenance and questionable shortcuts, even as several gathered with executive staff from the refinery in the evenings at tropically themed parties that flowed with booze and lavish meals.

Contamination, it turns out, was built into the design. As one investigator explained to me, "Every pipeline carrying a saleable product was built above ground. Every pipeline that carried waste products was installed below ground." Comprised of six miles of cast iron pipeline, some up to thirty inches in diameter, the entire waste stream was buried in the salty sand. The pipes started rusting almost immediately. In 1982, the refinery estimated 300,000 barrels of petrochemicals had leaked from these pipelines and formed a petrochemical slick some ten feet thick floating on top of the island's only aquifer. At one point, construction workers on the south shore stood back in surprise as a geyser of crude oil shot out of the hole they were digging. They thought they'd hit it big until the dismal reality of the situation became clear: they had tapped into a shockingly large plume of petrochemicals flowing from the refinery. An internal investigation in 2001 revealed 95 percent of waste-stream pipelines were leaking, and by 2005 the refinery concluded they were "deteriorated beyond repair."

Yet the refinery continued to operate as if nothing was amiss. By 2010, over one million barrels of oil had been extracted from the plume beneath the plant—an amount four times the size of the *Exxon Valdez* spill—yet the remediation of the plume was nowhere in sight. Carcinogenic vapors from petrochemicals are now readily detected in homes and neighborhoods along the south shore of St. Croix. When an EPA official went to St. Croix in 2011 to address the severity of what had been uncovered, they were shouted off the stage by residents furious over decades of quiet neglect.

Investigators also uncovered a history of shoddy practices that routinely sacrificed public health on the altar of operational ease and corporate returns. Workers told me stories of venting benzene under the cover of night on an island where residents still get their drinking water from cisterns and of flushing mercury down the drain into a bay still popular among local fishermen. Facing potentially record-breaking fines for this liable history of disregard, HOVENSA agreed to settle with the EPA in 2011. The refinery agreed to pay a $5.3 million fine and in lieu of penalties committed $700 million to extensive remediation, state-of-the-art pollution controls, and substantial investments in public health on St. Croix (including a cancer register to investigate residents' worst suspicions) (EPA 2011). At the time, this settlement was the largest on record for a refinery in the United States.

After finalizing the settlement, HOVENSA shut down and filed for bankruptcy in February 2012. This not only sidestepped its legal obligation to clean up its own mess, it also compelled draconian cuts to the territorial government budget. When the refinery shut its doors, 20 percent of the territory's annual budget disappeared in an instant. The closed refinery had "shaken the foundations" of St. Croix, the governor of the US Virgin Islands said at the time, forcing cuts that were nothing short of "catastrophic"(de Jongh 2012: 2, 4). Unemployment soon shot up to nearly 20 percent,c and energy costs skyrocketed (the refinery had long subsidized electricity and gasoline rates) as the state hemorrhaged governing capacity (USVI Source 2013). Crime rates on St. Croix rose substantially as theft and assault became commonplace. A UN report notes the US Virgin Islands now has the fourth highest homicide rate in the world (McCarthy 2018). One year out, the US Virgin Islands labor commissioner testified his surprise that there hadn't been a complete meltdown on St. Croix. "But," he added, "it has only been a year" (USVI Source 2013).

Catastrophe built on catastrophe. With St. Croix still in a tailspin, an unprecedented category 5 hurricane brushed up against it in 2017, causing considerable damage. Two weeks later, a second category 5 hurricane slammed directly into St. Croix, leaving nearly every building on the island in ruins and obliterating most public infrastructure. Ninety percent of all electrical transmission lines were destroyed. The back-to-back superstorms inflicted "widespread catastrophic damage," as NOAA put it, as uninsured damage exceeded $7 billion (Cangialosi, Latto, and Berg 2018). The hurricanes blew away roughly one in ten jobs on the island and hacked an already emaciated public purse in half (Austin 2020). Unemployment claims spiked to twice their previous high point: the closure of the refinery. The territorial government found itself downgraded and beyond bankrupt, unable to secure aid on a par with its dire need or to renegotiate its debt obligations. Four years on, congressionally allocated funds for recovery remained a fading promise in this Caribbean territory. The plight of St. Croix clarifies exactly what "colonial" means in the present.

In 2018, a year after one of the worst hurricane seasons in recorded history, Caribbean nations gathered to discuss climate resilience in the region. Many spoke of weaning themselves off fossil fuels and building green economies. The enthusiasm was clear: the Caribbean was poised to become the premier laboratory for redesigning societies beyond oil. Then the US Virgin Islands stepped on stage. Their plans for climate resiliency pivoted on one idea: restart the refinery. When pushed, officials spoke about the rising challenges and costs that climate change is bringing to the island with storms like Hurricanes Irma and Maria. How could the territorial government bear these costs without the refinery? The oil industry may be morally bankrupt and complicit in the coming catastrophe, but who else is still capable of paying the bills?

Since then, the territorial government has facilitated the sale of the refinery with generous tax breaks and promises to absolve the new owners of any responsibility for the legacy of contamination. The ongoing negotiations around restarting the refinery, according to recently disclosed internal EPA emails, "is receiving high visibility inside the beltway" in Washington, D.C., and is being used to showcase wider efforts to remake environmental protections as customer service for corporations (Hiar 2019). Assisting the new owners with "anything they need," wrote one senior EPA official, offers "a pilot for a broader customer liaison concept" in environmental enforcement (Hiar 2019). In this, the EPA seems intent on ensuring the new owners need not address the built-in flaws in the waste streams or the petrochemical contamination of the island. Indeed, recent EPA grants appear to use taxpayer funds to supplement the legal obligations of the refinery to install air monitoring equipment (EPA 2019). In 2020, the overriding priority of EPA was restarting one of the largest and dirtiest—indeed, most criminally negligent—oil refineries in the world. The EPA, one senior official wrote in 2017, "understands that restarting the operations at the former HOVENSA site would significantly benefit the economic health and well-being of the US Virgin Islands," something "especially important for the recovery of the US Virgin Islands in the aftermath of Hurricanes Irma and Maria" (1). The advice from President Donald Trump's EPA was clear: to survive the turbulence of climate change, we have no choice but to double down on the fiscal promise of oil.

When I visited St. Croix in the summer of 2019, the island still bore the impact of the superstorms. Tattered blue tarps fluttered atop buildings. Many former homes and businesses still lay in a mess of rubble and fallen trees, while whole sections of the main towns remained boarded up. Hurricane Maria destroyed eight of the thirteen public schools on St. Croix. Some four years later, reopening them remained a distant event. The sole hospital on the island now operates in patched-together units and temporary modules as the main building stands condemned, its roof leaking and the hallways overtaken by mold (Walker and Kanno-Youngs 2019). A replacement hospital has been promised for years, but construction is ever deferred. Medical care has been reduced to triage and basic care. Most everything else is referred to Puerto Rico or Florida. The island continues to lose residents; those who remain wait on promises still receding.

The refinery, however, is abuzz with activity. In 2019 I watched as piles of equipment twisted and broken in the storm were bulldozed to one side, while a main camp had been set up to house the one thousand workers from Texas and Louisiana tasked with getting the refinery operational again. Locals were promised jobs in the rebuilding project, but so far the only work to be found was providing menial services to the main camp in the evenings and on weekends. No one is entirely certain about the current levels of pollution around the refinery, as most of the environmental monitoring equipment stopped working when the hurricane hit.

In June 2019, local environmental and civic leaders gathered to discuss the urgent necessity of rebuilding St. Croix. One of the aims of this convening was to envision an energy system on St. Croix that was accountable to people and the earth. The organizers asked if I would join them to share my research on the refinery. We came together for two days at the still-shuttered Carambola Resort, with tarps stapled on damaged roofs and a beachside pool still full of debris. For many residents, the environmental injustice of the refinery and arrival of new superstorms were not unrelated events. They form a single continuum of fossil fueled disaster, a continuum that had to be broken if there was any chance of rebuilding with real hope. "From HOVENSA to Maria, there has been a plan to keep us down. We got to seek justice together."

"Why should we bear the burden for things others have profited from?" "Rebuilding is not enough, we must reclaim the land." One preacher offered an even longer history, noting "white supremacy and extractive capitalism are bound up together. They only see black people as something to use up and cast away."

"Oil sabotaged our island," a local farmer reflected on the last day. "And now it's up to us to set things right." Talking over the present plight for several days, the moment felt both desperate and pregnant with possibility. Again and again, someone would interrupt long pauses in discussions about the immensity of the challenge with the same refrain: "We need justice." And justice started with calling the fossil fuel industry to account for both the rampant contamination of the island and its stark vulnerability to the rising storms of planetary instability.

The negative ecologies of fossil fuels assail St. Croix from two sides: a catastrophic history of profitable neglect and a catastrophic future of climate instability. For state agencies and financial investors (and some strains of social theory), the answer to this conundrum is clear: it's only by turning our backs on the historical present that we can fully face up to the demands of the future. For the state, rebooting the refinery is the last gamble still offering winnings adequate to the great transformation now needed (without disrupting the neoliberal order of things). But doing so involves a technical baptism to wash the still-simmering history of toxic contamination from the official record. Such thinking advances new justification for substandard citizenship in places like St. Croix: to best prepare for climate change, we must absolve and subsidize the very industry that willingly led us into this crisis. Profits, not people, will save us in the end. Residents are having none of this nonsense. The disasters of toxicity and climate change may have very different temporal and spatial coordinates, but they share one liable author: the empire of oil. It is only by holding that empire accountable—by prosecuting the profiteers of destruction—that justice can be found and a society beyond oil begun. A climate crucible.

Conclusion

Negative Ecologies and the Discovery
of the Environment

Much has been made of the modern society's great divorce from nature, especially around the intellectual impoverishment of nonhuman vitality and the subsequent humanistic hamstringing of politics and ethics. Yet such scholarship strangely obscures the role of fossil fuels in this crisis. While it is widely recognized that fossil fuels instigated and may very well terminate industrial society's functional autonomy from the natural world, social research has largely turned a blind eye to the subsequent work of fossil fuels and the conditions of possibility they enable. Today, this is beginning to change. The groundbreaking work of Fernando Coronil (1997), Michael Watts (2005), Timothy Mitchell (2011), Robert Vitalis (2020), and others has shown how the shape and trajectory of our present carries the deep imprint of fossil fuels. While their work is instructive, the main thrust of their critiques rests on linking the material force of hydrocarbons to positive forms of capital and state power. But what of the disasters of fossil fuels?

The force of hydrocarbons is not fully expended in the moment of combustion, nor is it wholly transferred into accumulations of corporate profit or state violence. In cancerous bodies, asthmatic populations, scarred landscapes, and rising sea levels, fossil fuels disrupt the relationality of life on cellular and planetary scales. For many contemporary scholars of the oil industry, such destruction provides a useful backdrop en route to the main event: the accruals of power and profit in the empire of oil. Yet such destruction routinely surpasses the given ledgers of gain, whether such positive accounting pivots on the market valuation of the liable corporations or the technical ability to remediate the scene of the crime. Long obvious to frontline communities, the destruction unleashed by the oil industry grossly exceeds the reason of authority and profit. Medical journals now routinely report on the zones of injury that cohere in neighborhoods around refineries and petrochemical plants (Smargiassi et al. 2009; Deger et al. 2012), or even highways (Achakulwisut et al. 2019; Chakraborty 2022). This insight is becoming more commonplace as climate science and advocacy build a case for radically confronting the oil industry to triage a wounded planet. A recent essay in the stately journal *Science* laid out the straightforward case for "rapid decarbonization" (Rockström et al. 2017). From villages cut off from traditional abundance to contaminated municipal water supplies to distorted planetary systems, the physical destruction unloosed by the oil industry is becoming a more resonant experience than the fading promises of fueled progress, a more blatant truth than the rationale of accumulation, a more common ground than the common divisions of the contemporary. "Our addiction to fossil fuels is pushing humanity to the brink," UN Secretary General Antonio Guterres (UN 2021) recently said. "Enough of killing ourselves with carbon." Guterres demanded that we sound "a death knell for coal and fossil fuels before they destroy our planet."

The disasters of oil are more than a looming catastrophe; they are also a fractured history of our present. This book has explored the manifold oil disasters that stain our contemporary, from deepwater blowouts to extractive frontiers, from leaky refineries to petrochemical contamination. Each of these located disasters has seeded its own rippling creativity. As the oil industry assailed life in various quarters, it opened those now precarious conditions of life to new forms of understanding and responsibility. Over the past century and already spilling into our future, hydrocarbon harm has instigated the science and regulation of the environment. To a striking degree, the specific crisis the environment realized, the forms of understanding and responsibility it authorized, and the horizons of action and anticipation it routinized all bear the imprint of hydrocarbon afterlives. Whether by way of urban smog or petrochemical runoff or even oil spills, as the oil industry has unraveled the conditions of life, such destruction has also birthed new analytical techniques to monitor those conditions and new state authority to police them. Yet all too often we only grasp the disasters of oil through the resulting techniques and authorities.

We've treated the disasters of oil as not only secondary but as best understood by first placing them in the objectifying gaze of environmental science and policy. This not only dislocates and domesticates these disasters as marginal to the real story of crude oil, it also confuses the chronological and epistemic relation of disaster and the environment. As this book has argued throughout, the disasters of oil form the material ground upon which the environment is given scientific reality, even as the resulting science and its selective view become the official measure of those disasters. The chosen effects become the premier lens of the cause. Or, as Canguilhem ([1966] 1991: 243) writes, "The abnormal, while logically second, is existentially first" (see also Fortun 2001; Masco 2015). "Life rises to the consciousness and science of itself," writes Canguilhem (209), only through its own disruption. This book has explored this dynamic across a range of hydrocarbon installations in North America, describing how the manifold ecological disruptions of the oil industry continue to provoke the environment into being, largely by channeling the analytical legibility of those disruptions into the coordinates of thresholds and impact assessments. Yet even as the negative ecologies of the oil industry instigate the ongoing scientific and regulatory definition of the environment, those definitions never fully capture the materiality of the harm or its everyday experience. The negative ecologies of the oil industry are the fertile soil for the scientific ferment of the environment and, at the very same time, conclusive evidence of the environment's inability to truly reckon with the underlying crucible of vulnerable life. The environment works to prevent the negative ecologies of the oil industry from reaching a breaking point, from becoming a crisis. Or at least it did, until now.

It is this ongoing kilter of mastery and destruction that forms the ethnographic starting point, historical ground, and theoretical advance of this book.[1] Indeed, this dialectic between the negative ecologies of fossil fuels and the ongoing discovery of the environment is the pivot upon which this book revolves.

THE NEGATIVE ECOLOGIES OF FOSSIL FUELS

The world is coming undone at the hands of fossil fuels. What happens after the world's most prominent commodity now threatens the biochemical and meteorological conditions of life itself. While this destruction is not new, its widespread recognition is. Here, the ongoing substitution of hydrocarbon efficacy for coerced labor offers a tragic twist to the analytics of historical materialism. The telling tensions of our contemporary are ordered not only along the contradictions of production or power but also in accordance with the contractions of life, whether in rising rates of cancer or rising levels of seawater. Every human on Earth and most of the animals now host petrochemicals and other molecular traces of the fossil economy in their bodies. Hydrocarbons, in this reckoning, appear as a haunting gift, and perhaps it is Mauss more than Marx who offers the most exacting conceptualization of fossil fuels. When consumed, hydrocarbons do not disappear but can come to structure forms of obligation that may exceed the capacities of life itself. We may never run out of gas, but we are only just beginning to realize at what cost. In sharp and subtle ways, fossil fuels have assailed the underlying relationality of life and, by disrupting it, have opened vital ecologies to new forms of understanding and care. Prioritizing material fields of destruction and responsibility that cohere around fossil fuels—what I call the negative ecologies of oil—trips up historical genres of instrumental reason, neoliberal notions of marketable normality, and gainful economic measures, instead centering a viral devastation that only revolution can mend. While the rising prominence of the Anthropocene solicits rapt attention to the impending foreclosures heralded by hydrocarbon emissions, such a project frequently sidesteps the longer history of acknowledging and managing the disastrous qualities of fossil fuels (Bonneuil and Fressoz 2016). These concerns offer a generative site to take up questions of vital materiality and to reflect on the sometimes obscured relation of nonhuman agency to social justice. They also provide a new ethnographic and theoretical perch from which to understand the formation of the environment itself without taking its historical forms as normative.

THE DISCOVERY OF THE ENVIRONMENT

Over the course of the past century *environment* has shifted from erudite shorthand for *context'* to a proper noun worthy of its own governing agency in nearly every nation on Earth (whether that domain is called *environment, medio ambiente, mazingira, lingkungan,* or *huanjing*). This discussion stays close to the key problems and venues that brought the conditions of life into sharp analytic relief: the factory, the city, the nation, and now the planet. In many ways, the formation of the environment began in the factory. The newfound utility of fossil fuels and petrochemicals in World Wars I and II preceded any clear sense of the health risks they posed before (and after) the battlefield (Russel 2001). At once vital to the war and fraught with medical misgivings in their manufacture, these new ingredients of war became flashpoints between representatives of labor and representatives of industry. Federal officials in the United States, fearful of diminished wartime production, stepped in and asserted the power of science to defuse the historical antagonisms of labor and capital (e.g., Meeker 1920; Hamilton 1943). Responding, the rise of industrial hygiene came to format the interior spaces of industry as a technical field of scientific legibility, often around the maximum acceptable limits of hydrocarbon exhaust and petrochemical effluent for workers' health (Sellers 1997). Here, a fault line emerged in industrial production between an internal science of danger and an external science of gain, with the economy as a field of calculations taking forceful shape on one side of fossil fuels (Mitchell 2011) and the environment taking shape on the other.

The history of toxic exposure is also a history of analytic containers. In some ways, the emergence of the environment is the story of how the reach of petro-pollution came into focus outside the built mechanisms of control. In the 1960s and 1970s industries eluded regulation by designing bigger smokestacks or flushing waste downriver of the city or simply building new plants just beyond municipal or state jurisdiction. As lakes were declared dead, rivers caught fire, and mountains shorn of vegetation as rain turned acidic, the effects of diffuse petro-pollution became a rising public problem with apocalyptic overtones (e.g., Esposito 1970; Zwick 1971). Ecologists, turning their attention from the intricate balance of mountain ponds to the manufactured disarray of synthetic landscapes, began charting out how automobile exhaust or hydrochlorinated pesticides rippled through and unraveled the basic fabric of life (Carson 1962; Commoner 1967). As these insights were taken up in governance, the concept of the environment neatly joined this analytical field of manufactured disorder to the ethical object in need of saving. Consolidating previous concerns, the environment in need of federal protection since the 1970s has consisted of not only the monitored mediums of pollution within the nation but also the nation itself taken as a precarious system of life. As in the factory and in the city before, the state's enactment of a national environment did not register the historical inflections of class, race, and gender that were so often wrapped up in the toxic problems it purported to solve. Quietly orienting the state's forceful considerations as well as its averted gazes, the environment became both a vital object of contemporary politics and a technical limit to democratic practice. At sites ranging from leaky refineries to pipeline projects to oil spills to climate change, the disastrous properties of the oil industry continue to be at the forefront of innovations in environmental science and regulation. Though, as this book demonstrates over and over again, they do so less to confront that destruction than to provide the technical guidelines of assuring its smooth operation. It is also clear that this mastery was never complete, as the injuries of the oil industry have always refused easy objectification.

This book has explored the scientific and political dynamics at work around the negative ecologies of the oil industry and environmental protections across North America. Each chapter in this book has inhabited a place imperiled by crude oil. As the oil industry assails the vitality of the bayou or the boreal forest, the aquifer or the mangroves, that vitality becomes a field of sustained scientific scrutiny and a powerful moral counterpoint to the present destruction. Yet the resulting scientific and moral critiques do not, by default, stand against the oil industry. As I have shown, the orientation of scientific analysis and moral economy is in itself a new political field that forms around the negative ecologies of oil. The contrasts in these political fields across the chapters offer much to reflect upon. All the chapters draw attention to the sites imperiled by the oil industry and the ways the resulting damage opened them to new scientific appreciations. These scientific appreciations, moreover, did not retreat into the neutral world of facts but asserted themselves as powerful moral visions over and above the destruction that first formed them. But they did so in strikingly different ways: the early chapters show how that science can become ethical technologies that expand the reach of the state and even the oil companies, while the later chapters show how that science can become an insurgent solidarity to counter the abstract authority of states and corporations.

Far from the methodological assurances and moral insurance of safe distance, negative ecologies brought me closer to the world at hand in order to better see what is being done about it. The borderlands between mastery and destruction are tremendously generative, for regulatory science no less than for political protest. Not only are the negative ecologies of crude oil at the forefront of innovations in environmental science and policy, but their recognition leads many frontline communities to refuse offers of managed risk and instead call for the more fulsome condemnation of the oil industry itself, whether in occupied pipeline projects, sabotaged oil installations, or overrun hearings.

This book has advanced four lines of inquiry, which are particularly attentive to the ways in which (1) the material force of fossil fuels is not

FIGURE 13. Melting ice in Alaska. Photo by author.

fully expended in the moment of combustion but often comes to haunt life with sociochemical traces and attritional violence; (2) the scientific and regulatory definition of the environment does not precede the disasters of fossil fuels but often emerges from them; (3) the objectification of the environment glosses over embedded and embodied histories of harm; and (4) the empire of oil has not done away with nature but rather has unloosed new scientific and political desires for the natural world. Drawing these fields of inquiry and insight together, this book has advanced anthropological attention to the disastrous reach of fossil fuels. This book asks how a more deliberate ethnographic focus on the destructive capacities of fossil fuels might open new routes for social research to confront the political fields of risk and responsibility that are taking forceful shape around hydrocarbon infrastructure today. Among other things, this might help bring the distribution of uncertainty around hydrocarbon exposures into focus not as a momentary lapse in existing orders but as a generative disorder upon which social positions are realized and ranked anew or refused entirely as the negative solidarity of harm opens up a new commons. These themes and tensions might help us see the disastrous materiality of fossil fuels not as singular accidents condensed in time and space but as the fertile soil in which new political theologies of the injured life are taking root.

Acknowledgments

Friendship is the storehouse that provides for a journey like this. Old friends and new colleagues offered helpful directions when my thinking was adrift, a kind word when despair encroached, concrete support when it was most needed, and good company when all else failed. So many of my teachers and now my students have taken an interest in this project and asked me, in one way or another, to explain its merit, justify its labor, and perhaps do so with a good deal less jargon. Worthy measures, each of them, even as I fear I may never lay hold on all of them together. Many times I felt lost, as if I were hacking through a thicket, only to have someone suddenly point out the well-groomed trail just to my left. The deep well of generosity within this vocation of inquiry continues to astound: so many fellow travelers offered up spirited queries, pointed disagreements, and transformative suggestions. For this and so much more, special thanks go to Vanessa Agard-Jones, Anne Allison, Hector Amaya, Paul Andrews, Alessandro Angelini, Ben Anastas, Matteo Aria, Kate Aronoff, Fred Appel, Hannah Appel, Daniel Barber, Mara Benadusi, Amber Benezra, Mabel Berezin, Zeke Bernstein, Munirah Bishop, Mariah Blake, Maurice Block, Fred Block, Doug Bond, Soraya Boudia, Denise Brennan, Victor Breland, Molly Brown, Phil Brown, Imogen Bunting, Diego Caguenas, Maëlle Calandra, Brian Campion, Robin Celikates, Mike Cepek, Brian Chavez, Noah Coburn, Daniel Aldana Cohen, Phoebe Cohen, Ron Cohen, John Collins, Naja Collins, Alice Corbet, Rodrigo Cordero, Fernando Coronil, Gareth Dale, Mike Davis, Annabel Davis-Goff, Gregor Dobler, Freeman Dyson, Anne-Claire Defossez, Beshara Doumani, Andreas Eckert, Lynn Eden, Fran Edwards, Stacy Eisenstark, Judith Enck, Joel Fagerberg, Didier Fassin, Michael Fischer, Tom

Fels, Kadija Ferryman, Janet Foley, Andreas Folkers, Karolina Follis, Mike Fortun, Nancy Fraser, Elaine Gan, Ivy George, Frandelle Gerard, Deborah Gewertz, Brie Gettleson, Karen Gover, Lerone Grant, Nicolas Guilhot, Max Hanft, Angelique Haugerud, Maggie Hennefeld, Jesse Holcomb, Axel Honneth, Sophie Houdart, Kate Howard, John Hultgren, Murad Idris, Randi Irwin, Jon Isherwood, Rahel Jaeggi, Elio Jahaj, Deanna James, Sarah James, Ferne Johansson, Daniel Johnson, Corin Kaough, Daniel Karpowitz, Michael Kazin, Robin Kemkes, Max Kenner, Yasmine Khayyat, Doug Kiel, Stuart Kirsch, Scott Knowles, Nikolas Kosmatopoulos, Elbunit Kqiku, Greta Krippner, Stine Krøijer, Laura Kunreuther, Cedric Lam, Janelle Lamoreaux, Sam Lawson, Tess Lea, Sharon Lerner, Mandana Limbert, Peter Little, Claudio Lomnitz, Mark Longhurst, Yin Shao Loong, Nancy Lutkehaus, Rima Majed, Andreas Malm, Aldo Marchesi, Emily Martin, Andrew Mathews, Clara Mattei, David Meyers, Gregory Mingo, Dontie Mitchell, Charles McDonald, Carole McGranahan, Anne McNevin, Katiana Mentec, Sidney Mintz, Timothy Mitchell, Teresa Montoya, Amy Moran-Thomas, Andrea Muehlebach, Smoki Musaraj, Nate Olive, Juno Parrenas, Nishad Patnaik, Adriana Petryna, Elinor Phillips, Charlie Piot, Kari Polanyi-Levitt, Nicole Pontes, Vyajyanthi Rao, Ian Reid, Francisco Reinking, Thais Reis-Henrie, Liz Roberts, Susan Robinson, Doug Rogers, Janet Roitman, Jorja Rose, Olivia Rosenberg, Emily Sanders, Ritchie Savage, Suzana Sawyer, Tim Schroeder, Joan Scott, Jason Shafer, Nick Shapiro, Mimi Sheller, Dan Sherwood, Sommer Sibilly-Brown, Salma Siddique, Levi Silvers, Eric Smith, Cecile Stehrenberger, Ray Stevens, Emily Sogn, Grzegorz Sokol, Orin Starn, David Stoll, Ahwar Sultan, Yun Tang, Mick Taussig, Ute Tellmann, Dieter Thomä, Towanda Tilson, Alissa Trotz, Maria Urbas, Sarah Vaughn, Leilah Vevaina, Henrik Vigh, Bob Vitalis, Mieke Vrijmoet, Thuc Linh Nguyen Vu, Greta Wagner, Sophie Wahnich, Casey Wait, Jesse Walsh, Kaedra Walsh, Michael Watts, Mike Weaver, Aila West, Gisa Weszkalnys, Charles Whitcroft, Hylton White, Jessica Winegar, Kerry Woods, Rebecca Woods, Deborah Yashar, Ahmad Yassir, Qiaoyun Zhang, and Yuan Zhang.

I would be remiss to neglect the institutions of learning that have stood by me, both run on the same mix of shoestring budgets, progressive commitments, and a real humanity at their core. The New School for Social Research proved a transformative home for my graduate studies, a place where the study of politics is never far from its spirited practice. Small, quirky, and always punching above its weight, Bennington College offered up a feast of interdisciplinary camaraderie, complete intellectual and pedagogical freedom, and earnest students always willing to see where a new question or insight might take us. Material support from both institutions, along with the American Council of Learned Societies (ACLS), the National Science Foundation (NSF), and Wenner Gren, proved instrumental to funding the research that poured the foundation of this book.

The book itself, and the labor of stitching together the threads of various inquiries, first found its way during a magical year at the Institute for Advanced Study. Stacy, Naja, and Francisco from the University of California Press offered deft hands that shepherded a fledgling manuscript into this much improved book. Sharon Langworthy's copyediting made key investments in the clarity of this book. Daniel Baltrá's portraits of the oil spill in the Gulf of Mexico floored me when I first saw them. They conveyed a truth I was struggling to put to words: oil spills draw us into seeing the natural world in profoundly new ways and render our precarity most knowable in its passing. I am truly honored that one of his images graces the cover.

In a moment of profound lack, Karen and Randy Whitaker offered a double-wide on a hilltop; its stunning view allowed me to see the horizon once more. Elizabeth Coleman made an early bet and blasted open a pathway to employment where none previously existed. With arms open to the tremendous need of this world and an abiding commitment to do something about it, I could not have asked for a better boss than Susan Sgorbati. At a difficult crossroads, Joseph Masco offered honest encouragement and real assistance, for which I remain grateful. When I first met Kim Fortun, this project was a set of images and encounters I could not shake. Repeatedly, Kim invited me to work out the connections in the generous and generative company she keeps.

By a fortuitous turn of events, I ended up in the first class Ann Stoler taught at the New School for Social Research. I knew almost immediately, Ann's version of anthropology was where I would take root as a scholar. Ann also provided a model for teaching and critical engagement that I still labor to live up to. Lucas Bessire has been a sounding board for this book since its earliest inkling. So much of the questions and commitments that inform this book took shape in our conversations, and our friendship remains a dear compass.

My stepfather, Bryan Lewis, one told me how remarkable it was to find a job you truly enjoy. That he considers me fortunate to do what I love means quite a bit. My mom, Marion, has been a constant source of inspiration. Her undaunted spirit and sense of adventure have proved vital tools that helped me find my way. She believed in my ability even when everyone else pointed out the ample evidence to the contrary. From Brooklyn to Vermont, Carrie Gettmann helped provide a home at every iteration of this book. It could not have been written without her steadfast support. With irreverence and radiance, Munira Khayyat illuminates the landscapes now in need of tending and the horizons now standing within reach. Najmati al janubiya. William's keen eye for the small surprises of the world have kept me from moving too fast. Meredith's rambunctious joy reminds me of everything else that matters. Together, they are the world I hold most dear.

Notes

Notes

CHAPTER I. ENVIRONMENT: A DISASTROUS HISTORY
OF THE HYDROCARBON PRESENT

1. For the first half of the twentieth century, "environmentalism" was a distinguished current of theory within US social science, one emphasizing contextual explanation in scholarship. As the US social sciences' first translation of Auguste Comte's *milieu* and Wilhelm Dilthey's *Umwelt*, the term *environmentalism* foregrounded the influence of surroundings in understanding of social phenomena. Environmentalism was not a substantive problem, a regulatory project, or a stable scientific field; it was a generic privileging of context in social explanation. As "environment" became the premier diagnosis of the crisis of life in the 1970s, the intellectual history of the term fell by the wayside as the operational coordinates of the environment came to reign supreme. This etymology is explained later in this chapter (see note 45).

2. James Hansen's 1988 testimony is widely celebrated as a watershed event, introducing climate change to federal governance (*Hearing on Greenhouse Effect and Global Climate Change* 1988). Ten years earlier, the second EPA administrator, Russel Train, who was appointed by Richard Nixon, wrote in *Science*: "There is growing scientific concern over the buildup of atmospheric carbon dioxide from the combustion of fossil fuels with potentially significant impacts on global temperature and climate. All of this suggests that coal (sometimes described as America's energy 'ace in the hole') may be a very uncertain foundation upon which to base long-term energy policy. [. . .] The world will have to turn away from fossil fuels long before usable coal reserves are exhausted" (1978: 322). Indeed, climate change was also on the docket in

congressional hearings about the National Environmental Policy in 1968 and 1969, where "the contrary possibilities of rising world temperatures as a result of carbon dioxide build-up or falling temperatures as a result of smog and jet contrails" were introduced as possible fields of responsibility for the governance of the environment (*Hearing on Environmental Quality* 1969: 15486). The White House commissioned a report on pollution in 1965. In a chapter entitled "Carbon Dioxide from Fossil Fuels: The Invisible Pollutant," the report stated "the data show, clearly and conclusively" that global CO_2 levels were rising in tandem with fossil fuel consumption (White House 1965: 116). Noting "the extraordinary economic and human importance of climate," the report described the unregulated burning of fossil fuels as "a vast geophysical experiment" (126) that needed to be addressed to "counteract" the planetary effects of burning fossil fuels (127). It is not the recent discovery of the direct connection of fossil fuels and climate change that should hold our attention but the ease with which that long-standing fact has been forgotten.

3. "The environment has a history," writes Paul Warde, Libby Robin, and Sverker Sörlin (2018: 180) in their seminal intellectual biography of the environment. Yannick Mahrane, Marianna Fenzi, Céline Pessis, and Christophe Bonneuil (2012:xiv) also insist the concept of the environment has a history in their incisive sketch of the global convergences that forged the "environment as an international political object." These recent historical projects stake out a bold reappraisal of the environment by asking the most straightforward of questions: how did such a pedantic term like the environment gain such public prominence? For far too long, the environment was something to be allied with, not analyzed itself; something to explain with, not something to explain. The environment was more an orientation in scholarship (and activism) than an object of scholarship (ie, environmental anthropology, not an anthropology of the environment). The recent historical attention to the formation of the environment itself traces out the uneven intellectual terrain of the concept from the late 1940s to the 1970s, its introduction of the precarious future to science and statecraft, its global spread and planetary pretensions, and the prominence of the United States. My provisional account of the environment follows in the footsteps of these groundbreaking histories, while emphasizing how the petro-materiality of American power helped instigate, instantiate, and institutionalize the environment as an object of science and politics.

4. Some scholars have equated the rise of environmentalism with middle-class aesthetics, whether by setting aside certain ideals of nature for weekend consumption or the suburban bastions of not in my backyard activism (Radkau 2014; Sellers 2012). Others have located the true birth of environmentalism in the way peasant land ethics came to protest the arrival of extractive industries (Taylor 1995; Guha 2000; Martinez-Alier 2002). My point here is that the field this social movement organized around—the environment—first gained operable form primarily in relation to the negative ecologies of contemporary power. This book presumes that many of the social movements that have cohered around the environment in a variety of local, national, and international contexts have long tripped up and exceeded the technical constitution of the category, even as the state-backed objectification of the environment

has animated the analytical and ethical justification of those movements. More recently, historians have turned attention to the environment itself (Warde, Robin, and Sörlin 2018), yet they do so primarily through the history of ideas set apart from the material problems, geopolitical struggles, corporate agendas, and governing logics that instantiated the environment.

5. Perry Anderson, in an interview with Raymond Williams, defined structures of feeling as "the field of contradiction between consciously held ideology and emergent experience" (1979: 167). Perhaps, and Williams certainly wrote of his intention to brush aside overly formalist explanations that seem to abstract social life so that critical scholarship, like literature, might grasp "meanings and values as they are actively lived" in the taut gap between the official version and everyday life (1977: 132). But this formulation may shortchange the creative materialism at the very core of Williams's project. In response to Anderson, Williams pushed against the "rabid idealism on the left in the sixties and seventies" (1979: 167) and the manner in which "great blockbuster words like experience" (1979: 168) all too quickly fence off materialist fields of inquiry and analysis, holding all manner of things at bay. In a discussion long neglected in the subsequent popularity of the term, Williams recentered structures of feeling on specific material ruptures that find themselves, as he had written in *Marxism and Literature*, "at the very edge of semantic availability" (1977: 134). Structures of feeling, Williams wrote, "initially form as a certain kind of disturbance or unease, a particular type of tension, for which if you stand back or recall them you can sometimes find a referent" (1977: 134; 1979: 168). One thinks of the novelty of country estates in *Country and the City* (1973), the arrival of railway in the Welsh Trilogy, and perhaps ecological crisis in his later reflections on nature. For Williams, these disruptive material referents at the core of structure of feelings are always tethered to the expanding history of capitalism. With DDT and strontium 90 I wonder if we might allow for a disruptive materialism in excess of capitalism and see a sprawling structure of feeling taking shape around the negative ecologies of hydrocarbon prosperity and thermonuclear statecraft.

6. The turn to supplement the sun's seasonal energy with geological storehouses of hydrocarbon energy—first with coal and then crude oil—conspired with new logics of profit and practices of rule to spark uneven transformations within human society. While scholars continue to debate what part of this transformation was due to fossil fuels and what part was due to capitalism (Ruddiman 2005; Malm 2016), it is clear that fossil fuels profoundly shaped the texture and trajectory of the modern world. Stored deposits of hydrocarbon energy helped fuel mechanized production in factories, concentrated misery in cities, and helped expand the geography of markets while vastly increasing the scale of what could be transported and to where. While the narrow conduits of coal provided the technical basis of broad social democracy, the more imperial and flexible conduits of crude oil provided its limits (Mitchell 2011). The "privatized mobility" of the automobile, as Raymond Williams put it, transformed the city, exiling the vibrant public life of urban streets to private interiors (Norton 2011). The car also underwrote the construction of the suburbs, with new petrochemical lifestyles and plastic subjectivity (Huber 2013).

As Max Weber ([1990] 2002: 123) once noted with a qualifier on social theory that we have yet to fully reckon with, perhaps this iron cage will hold "until the last ton of fossil fuel is burnt to ashes." Rather than relitigating this robust and still unfolding debate, my point here is to simply note how the consequences— the negative excess of fossil fuels and nuclear weapons— exceeded the existing frameworks of capitalism and the state. While we may require a thorough theory of capitalism and the state to understand how fossil fuels and nuclear weapons were put into play, their effects soon exceeded the analytic capacity of those theories.

7. Theodor Adorno ([1966] 2007: 5) insisted that things "do not go into their concepts without leaving a remainder," and a new critique might be born from attention to that haunting excess of commodification. With attention to peasant frontiers around the world (Wolf 1969, 1982; Taussig 1980, 1991, 2018; Ong 1987; Stoler 1995), the dark shadows cast by the commodity have long been at the forefront of anthropological critiques of capitalism. Today, negativity, as an experience and a condition, is moving from anthropological histories of commodification into the wider landscapes of violence and dispossession (Stoler 2013; Masco 2015; Bessire 2021). Far from a lack, Gaston Gordillo (2014) demonstrates the generativity of the negative in his adept excavation and narration of the ruins that dot the Gran Chaco. Lucas Bessire (2014) turns to "negative imminence" to describe the haunting creativity of the Ayoreo who refuse colonial and humanitarian grids of recognition while insisting of present-tense difference. Yael Navaro (2020) calls on anthropology to take up a "negative methodology" more attentive to the gaps and hollows that accompany social violence, and what work unfolds in those absences. The negative, as existential approximations of losses that defy the reason of gain, remains uniquely accessible to ethnographic description and at the forefront of one genre of ethnographic critique. Defying the positivist methodologies of sister social sciences, anthropology can reflect on the existential and historical dimensions of negativity without first presuming its technical resolution as the baseline of apprehension.

8. Ecology also has a distinguished place within the history of anthropology. In the 1950s, ecology became a prominent frame for situating the practice of culture within the allowances of regional landscapes (Steward 1955; Netting 1977; Rappaport 1968). More recently, ecology has come back into broad anthropological currency as a window to apprehend multi-species ways of relating that surpass the humanist format of modernity, and its impending collapse (Kirksey 2015; Tsing 2017; Stoetzer 2018; Morimoto 2022). We might say that while anthropology previously emphasized the gravity ecology places on present life, more recent monographs emphasize the escape velocity ecology inspires for future life. Within the prose of ethnography, ecology is shifting from a limit on social possibility to a launchpad of social possibility.

9. While they were by no means the only discordant things to capture broad attention, DDT and strontium 90 became the premier examples of the ecological underside of petro-prosperity and atomic might. A number of studies in the 1960s indicated that pollution had penetrated into the deepest recesses of human biology: debilitating levels of lead from gasoline emissions were

readily found in blood samples in cities across the United States; autopsies in Pittsburgh, Montreal, and New York City revealed roughly half of the bodies examined had tumors from asbestos in their lungs (cited in Esposito 1970: 15); carbon monoxide from automobiles and dissolved nitrates from fertilizers both inhibited the human body's ability to absorb oxygen, artificially stunting the functions of various vital organs (Carr 1965; NAS 1969); a radioactive isotope of strontium, released into the atmosphere in nuclear tests, rather quickly found its way into human bone marrow and developing fetuses (Kulp, Ecklemann, and Schulert 1959); and hydro-chlorinated pesticides like DDT were rapidly accumulating in the fatty tissue of nearly every human on Earth (White 1964; Rudd 1964). See Fowler (1960); Commoner (1967), Rudd (1964), and above all others, Carson (1962).

10. Strontium 90 accumulated at alarming levels in the fast-growing bones of breastfeeding infants and developing fetuses, a process that the rising incidence of leukemia in US children during the 1960s and 1970s was attributed to (Newcombe 1957; Eckelmann, Kulp, and Schulert 1958; Kulp, Schulert, and Hodges 1959, 1960; Kulp and Schulert 1962). These discovered linkages led to the widespread collection and study of baby teeth as a crucial indicator of the spread and density of radioactive fallout in human bodies (Reiss 1961). As Joseph Masco (2016) demonstrates, this most unnatural injection of radioactivity has now settled into the comforts of natural background level, where its potency winds down over geological timescales but is no longer seen as an event.

11. Although set in motion by Cold War bluster, the belligerence of development, and blithe suburban consumption, the disruptive afterlives of DDT and strontium 90 exceeded the geographies of those projects and their analytical jurisdiction. Empire, capitalism, and the state may be required references in understanding how DDT and strontium 90 first were set loose in the world, but once unloosed they required a new science to fully grasp.

12. This "Age of Fallout," as anthropologist Joseph Masco (2015, 2021) has aptly named it, introduced a new kind of "invisible injury" that heralded profound ruptures of life but unfolded on scales and temporalities just beyond the registers of human experience, whose very visibility was deeply reliant on national security and yet always exceeded the operational capacities of that increasingly dated political form (whose very "datedness" was in no small measure a result of problems like this), and whose analytic and affective texture prompted political subjects to remake themselves in relation to a future-oriented form of historical reckoning only "made visible in negative outcomes."

13. Such a conception stands at odds with critical appraisals of crude oil (and nuclear weapons). Distilling a popular current of social research and scholarly critique, Matthew Huber (2013: 3) has written of oil: "My wager is that *it* doesn't do anything. Oil has no inherent power outside the social and political relations that produce it as such a 'vital' resource." This approach to the embedded power of oil has proved critically incisive in many ethnographies of oil, whether by locating the material question of oil in the field of Indigenous politics (Sawyer 2004; Cepek 2018), the neoliberal family and consumer citizenship (Shever 2012; Huber 2013), the cultural reach of state power (Apter 2005; Rogers 2015), the temporality of development (Limbert 2010; Weszkalnys

2014), or the infrastructural life of extractive capitalism (Barry 2013; Appel 2019). I have no disagreement with this prerogative to attend to the embedded social constitution of oil's power – indeed, I continue to learn from and teach with this accomplished body of scholarship – as long as it is lodged in the world *before* combustion. In the aftermath of combustion, fossil fuels gain a force that is neither empirically nor theoretically contained within the labored dimensions of the commodity nor in the social format of neoliberalism, that actually does seem to be rooted in the agentive physicality and functional autonomy of the thing. Capitalism and the state are requisite histories in any serious account of how fossil fuels and nuclear bombs come to exert such influence in the contemporary world. But what happens after the commodity, after the weapon, stands in ecological excess of those histories. The aftermath unfolds with a causal force that will continue inflicting disruptions to life regardless of the subsequent rise or fall of capitalism or American empire or humanity itself.

14. Nuclear fallout grossly exceeded the intended violence of the weapon. The World Health Organization commissioned a report in 1983 that aimed to measure the health effects of nuclear war. The report found that in an "all-out nuclear war" 1.1 billion people would die within the first few hours of the war, and another 1.1 billion would be permanently disfigured, while every major city in Europe, the United States, and the Soviet Union would be rendered uninhabitable for generations (WHO 1983). As Carl Sagan responded to this alarming report in *Parade Magazine* in 1983, "Unfortunately the real situation would be much worse" (4). Coining the term "nuclear winter," Sagan summarized the emerging atmospheric science and planetary modeling of nuclear war. The explosive blasts merely provide the opening skirmish of the global tsunami of destruction. Through supercharged firestorms that would consume entire landscapes, smoke and dust suspended in the stratosphere that would obscure the sun for years, and toxic fallout that would contaminate prime agricultural land, nuclear war would render the entire planet burned, dark, cold, radioactive, and inhospitable to life. "Except for fools and madmen, everyone knows that nuclear war would be an unprecedented catastrophe," Sagan wrote (1983: 4, 7), yet "fools and madmen do exist, and sometimes rise to power." The analytical grasp of this unbound destruction of nuclear weapons initially moved in two opposite directions. Within the state, a new expertise arose focused on how to survive it domestically—an expertise that came to privilege vital infrastructure over people and redesigned much of American life accordingly ("reflexive biopolitics," as Collier and Lakoff 2021 have perceptively described it)—and how to weaponize it externally. Realizing the ecological shockwaves were far more destructive than the initial blast, many military scientists worked to better aim and amplify fallout as the real weapon of atomic explosions (Hamblin 2013). Recoiling in shock over the same emerging scientific picture and planetary dimensions of fallout, many academic scientists became convinced that rampant militarization of synthetic force might actually snuff out life on Earth (Commoner 1971; Crutzen and Birks 1982; Grover and Harwell 1985). They built a compelling scientific case for radical structural change.. As Donald Worster (1994: 339) has written of this moment, "For the first time in some two million years of human history, there existed a force capable of destroying the entire fabric of life on the planet."

15. The theory of lively materiality offered by the ecologies and outrage over the atomic bomb remains unexplored. There are, however, a few provocative openings. "The Bomb is, after all, something more than an inert Thing," wrote E. P. Thompson (1982: 4). Nuclear weapons are an unprecedented menace, they are embedded in highly automated systems programmed to override human reflexivity, and they are a force that once unleashed quite soberly promises totalizing extermination. A decade or more before such phrasing entered the lexicon of critical theory, Thompson concluded: "Weapons, it turns out, are political agents" (7).

16. While they were by no means the only discordant things to capture broad attention, DDT and strontium 90 became the premier examples of the dark ecological underside of contemporary imperialism. A number of studies in the 1960s indicated that leaded gasoline, carbon monoxide, sulfur dioxide, iodine 131, cesium 137, and other forms of contemporary pollution also had penetrated into the deepest recesses of human biology (LI 1969; Carr 1965; NAS 1969).

17. Rachel Carson (1962: 188) wrote: "The new environmental health problems are multiple—created by radiation in all its forms, born of the never-ending stream of chemicals of which pesticides are a part, chemicals now pervading the world in which we live, acting upon us directly and indirectly, separately and collectively. Their presence casts a shadow that is no less ominous because it is formless and obscure, no less frightening because it is simply impossible to predict the effects of lifetime exposure to chemical and physical agents that are not part of the biological experience of man."

18. This petrochemical-hybrid complex neatly packaged and exported in the Green Revolution should also be seen as counterinsurgency by other means. As tractors and petrochemicals helped displace family farms and consolidate agriculture at industrial scales, we should recognize that the synthetic productivity of the Green Revolution made the countryside of the Global South newly dependent on American agribusiness and petrochemical firms while simultaneously emptying the countryside of pesky peasants and their penchant for agrarian revolution in the twentieth century (Shiva 1989; Davis 2006). Deriving nitrogen from natural gas actually "makes fertilizer production the largest energy input into US industrial agriculture," Moore and Patel (2018: 174) write. In so doing, the Haber-Bosch Process has allowed "meatification" of global diets (Moore and Patel 2018: 174).

19. This emphasis on synthetic might is neither to dismiss nor to downplay the ongoing colonial occupation of Native lands; the military invasions of Afghanistan, Bosnia, Grenada, Guam, Iraq, Korea, Kuwait, Panama, Marshall Islands, Vietnam, and elsewhere; the clandestine terrorism of the CIA across Africa, Latin America, the Middle East, and South Asia; and the financial terrorism of US-backed structural adjustment across the Global South. Rather, it is to emphasize a primary material pedestal from which American empire came to flex its might in the twentieth century, one that helped give rise to the notion—for adherents as much as critics—that American imperialism exemplified a different kind of empire.

20. As Rachel Carson's editor wrote to her about the manuscript that became *Silent Spring*, the proverbial lightbulb flipped on with the recognition of

"fallout" as the perfect concept to hold together the analogy of atomic bombs and petrochemicals (Radkau 2014: 75).

21. For example, Paul Crutzen, who coined the term "Anthropocene," rose to prominence for his analysis of the "climactic effects of nuclear war" (1987) before becoming one of the premier atmospheric chemists studying the climate effects of rampant consumption of fossil fuels and petrochemicals.

22. Prescient of our contemporary, the recognition of negative ecology brought the hard-nosed sciences into practical demands for revolution, while those presumably schooled in radical social change—the progressive social sciences—seemed to drift either into liberal compliance or esoteric irrelevance to the more material questions of crisis at hand.

23. As an analytic concept, "structures of feelings" insists upon change. "Perhaps the dead can be reduced to fixed forms," Raymond Williams wrote, "but the living will not." The present is always at the cusp of a confluence of archaic forms, the existing order, and new things. Standing within a quickening current without an agreed upon course, the present unfolds as a passionate affair not yet weighed down in the formal dress of "definition, classification, or rationalization" (1977: 132).

24. Raymond Williams privileged art and literature for their early depictions of these ruptures rippling out across a society before either the state or the sciences had a name for such changes. Such novels or works of art, Williams wrote, "produce a shock of recognition" among their contemporaries (1979: 164). "What must be happening on these occasions is that an experience which is really very wide suddenly finds a semantic figure which articulates it" (1979: 164). Surely literature has no monopoly on such prescience. While some scholars have commented on the astounding popularity of *Silent Spring* by referencing the wider anxieties of the Cold War and racial strife, I might also insist on a more literal reading. *Silent Spring* provided a new language of empirical and moral discontent around the suspicion that the massive chemical expenditures fueling headlong prosperity might also be undermining the possibility of life within such relentless progress. Science introduced a new vocabulary to grasp the massive changes underway.

25. While initially many of these new threats showed the privilege of the suburbs to be no shield against toxicity, the proffered solution frequently rested precisely on securing just that privilege. If DDT and strontium 90 helped first illuminate these concerns, today toxins like atrazine, dioxin, PFAS, plastics, and above all CO_2 do similar work. They act as tracers that illuminate the poverty of our existing structures for fostering life and illustrate how interwoven life is even as they draw its induced endpoint into focus. And they pose the ethical dilemma: there is no solution that rests on barricaded islands that disregard the remainder of humanity. So can we recognize and build a politics adequate to the commonality of need, or will we continue following the promise of walls until everything falls apart?

26. After World War II, abundant reserves of crude oil in California, Oklahoma, and Texas provided for America's growing addiction to cheap energy. Yet after huge discoveries of oil were made in the Middle East in the 1940s and 1950s, President Dwight Eisenhower launched the Mandatory Oil Import

Program (MOIP). At first, these overseas discoveries flooded the domestic market with cheap Middle Eastern oil, collapsing the domestic prices of crude. It was a market reality that threatened to bankrupt US oil companies, many of which were invested heavily in aging domestic reserves that required costly interventions to keep them producing. Imposing strict quotas for oil imports at around 12 percent of total domestic consumption, from 1959 to 1973 MOIP aimed to minimize dependence on foreign oil while propping up the solvency of oil extraction in the United States. With the resulting public investment in their profitability, those oil companies pivoted to an expanded presence in foreign oil. Between 1940 and 1967, "US companies increased their control of Middle Eastern oil reserves from 10% to close to 60%" ("Editorial" 2002: 2). When the next energy crisis hit—first the depletion of domestic reserves and then the OPEC embargo of 1973 and 1979—those same oil companies lobbied to do away with import controls and used the pretext of the crisis to more effectively deregulate the oil industry and more effectively decouple its vitality to the economy from its destruction of the environment. This also led to a significant consolidation of the industry: by 1970, 70 percent of the oil consumed in the United States was provided by just twenty oil companies, most of whom were more firmly organizing themselves as transnational corporations (Jacobs 2016).

27. Some scholars suggest deeper connections between the new militarized US expeditions around the world and the rise of environmental consciousness. For at this moment Americans "began to see the whole planet—the Earth itself—as in some ways American" (Robertson 2008: 584). The environmental mantra "think globally" coincided with US interventions at that very scale.

28. Energy, of course, is only one of the contradictions that gave rise to this formation. This image of prosperity was also underwritten by racial exclusions, tyrannical gender norms, and growing military investments in neocolonial escapades to secure bargain-bin prices for natural resources under a new ideology of development.

29. Much is made about the negative externalities—how our premier economic models struggle to account for the harms that are part and parcel of the gains produced, whether in pollution, toxic waste, or health impacts—yet it might be recalled that this externalization was itself an achievement, and much of this effort centers on making these externalizations measurable within the logics of gain.

30. As a 1971 *Readers Digest* supplement, "What Are We Doing about Our Environment?," makes clear, car manufacturers, the petrochemical industry, and others had no difficulty incorporating the idea of the environment into their operations.

31. Naming the resulting methodical and systematic attention to air "the climatology of combat," Sloterdijk (2009: 19, 30) argues that this was "the first science to provide the 20th century with its identity papers."

32. This relation did not go unnoticed. Reviewing the dangers of fumes in munitions factories, dean of Harvard Medical School David Edsall (1918: 199) commented, "The main effects of these fumes are much the same as those chief gases used in warfare."

33. Alice Hamilton decided to pursue an education in medicine "because as a doctor I could go anywhere I pleased" (1943: 38). Having studied bacteriology and pathology at the University of Michigan, the Universität München, and the Institut Pasteur, Hamilton was asked by the governor of Illinois to investigate rising incidents of "occupational diseases" in Chicago factories in 1910. "It was pioneering exploration of an unknown field" (1943: 121). This experience led to her being appointed to head the Bureau of Labor Statistics (BLS) investigations of factory conditions nationwide during World War I. After the war, she was appointed professor in the new department of industrial medicine at Harvard in 1919. The *New York Times* headline ran: "A Woman on Harvard Faculty—The Citadel Has Fallen." Hamilton was the first female professor at Harvard, an honor awarded her with the strict provisions that she never set foot in the faculty lounge, never attend football games, and refrain from public participation in the graduation ceremony. Choosing to limit her time in Cambridge to one semester a year, Hamilton lived most of her life at Hull House with Jane Addams in Chicago, where the likes of Eugene Debs, Clarence Darrow, John Dewey, and Upton Sinclair regularly crossed her path. The "dangerous trades" that framed her autobiography included older industries that used mercury, lead, and radium alongside new production processes for leaded gasoline, petrochemicals, and chemical munitions.

34. Note the clinical focus on the workers' body in her investigations: "A boy of 18 was set to clean out eighteen to twenty amastol tanks during the heat of August. He worked only three days and then went to the doctor, deeply jaundiced, the whites of his eyes a bright yellow. He was vomiting, had diarrhea, pains in the abdomen, weakness of the knees, and a greenish tinge in the urine" (Hamilton 1919: 257). "Inhaling carbo-hydrogen gases . . . [resulted in] stupor, headaches, inclination to cough, acid eructations, a buzzing noise in the ears, vertigo, intoxication, tremor, and convulsions" ("Industrial Poisons and Diseases" 1918: 187). "A pale, coated tongue, slight loss of appetite, vomiting, and sometimes slight jaundice, and a little albumen in the urine," wrote Alice Hamilton (1918: 243). "The workers show marked lassitude."

35. Alice Hamilton (1919: 249) noted that factory physicians "did nothing to prevent sickness, and not very much in the way of treatment." If these physicians thought a worker was poisoned, "they promptly discharged him in order that he might not become a charge on the company" (249). In response to such practices, several unions organized their own health clinics to police and remedy the toxic exposures of the workplace. With the rise of industrial hygiene, however, most union clinics bequeathed such medical responsibility to the new government agencies (Rosner and Markowitz 1989).

36. Stabilizing the object of their science legitimated and oriented their analytical operations as science. In this, industrial hygienists no longer needed to conduct medical fieldwork inside the factory. Research on working conditions took the form of brief inspections, a series of measurements, and a checklist of required safety features. Once the dangers were fixed as discrete things, calculating the risks of the factory became a technical operation.

37. As Senator Edmund Muskie (D-Maine) and President Richard Nixon squared off over the environment in anticipation of the 1972 presidential

election, both were convinced the environment provided an effective route to a new winning constituency. Neither strategy was without cynicism. While Muskie saw the environment as the next unifying issue for the Democratic Party (neatly sidestepping civil rights), Nixon hoped the environment might deflect attention from the war in Vietnam among younger voters. Ultimately, the upstart McGovern bested front-runner Muskie (after Nixon's henchmen forged a letter in which Muskie seemed to be mocking ethnic communities, the key to the Democratic coalition) in the Democratic primary and Nixon walloped McGovern in one of the most lopsided presidential elections in American history (with McGovern winning just one state, Massachusetts).

38. The independence of the EPA was a crucial achievement. Many Republicans, especially those from oil-producing states, urged Nixon to create a new agency that would balance natural resource development (i.e., extraction) and environmental protection. As Walter Hickel (1971: 243), tireless promoter of Alaskan oil and secretary of the interior under Nixon, explained, "I reasoned that it was self-defeating to separate resource development from environmental protection."

39. Here, the natural world, specifically local ecosystems and planetary systems, was recognized not through what *homo economicus* might make of it but through it independent capacity to sustain life. This recognition that the natural world has agency beyond the human belies the more breathless accounts of posthumanism and the chronologies they rest on, as does the manner in which these early insights fed into expansions of state power.

40. When challenged by a colleague over how the views of the natural world could ever be known, Christopher Stone replied: "If you listen very, very closely, a tree will make the exact same sounds as a corporation" (cited in Mosk 1976: 231).

41. This emphasis on vulnerability—as an emergent mood and a specific scientific problem that cohered around the negative ecologies of fossil fuels—as the foundation upon which the environment emerged stands in sharp contrast with prominent theories of the environment that emphasize scarcity. Whether as a continuation of the conservation ethos and policies (Hays 1959, 1989, 2000; Nash 1990) or as the institutional failures of the liberal state to keep up with the treadmill of production (Schnaiberg 1980), scarcity has frequently been placed at the center of the historical formation of the environment. The argument here is that the environment is not so much an extension of previous economic logics of scarcity into new ecological realms as the rise of a new scientific and regulatory regime around vulnerability. The rise of vulnerability comes to index not so much a quantitative lack as an emerging qualitative condition calling for either revolution or a new regime of security. For insightful scholarship that foregrounds the rise of vulnerability, see Collier and Lakoff (2021) and Vaughn (2022).

42. The Council of Environmental Quality was explicitly modeled on and sought to counterbalance the Council of Economic Advisors.

43. If the Foucauldian overtones of Caldwell seem striking, they may not be entirely out of place. Before he left Indiana to draft NEPA, Lynton Caldwell wrote a short reflection on the new science required to hold together the insights

of research, advocacy, and regulation that the crisis of life brought into lively intersection. Noting the "inadequacy of conventional political mechanisms to deal with the problems of the new age of biology"—that is, biology as the study not of life but of altered life—called for a "new machinery in governance" (Caldwell 1964: 28, 29). Such governance would coordinate the emerging science of life's precarity and the moral discontent it gave rise to into an effective federal governance of the natural world. At that "meeting point of science, ethics, and politics," Caldwell saw a pressing need for a new discipline to hold everything together, a discipline he named "biopolitics." Biopolitics, he argued, could finally turn the tide and counter the biological perversions of the hydrocarbon and radioactive present with a more informed, more unified, and more forceful defense of the precarious conditions of life itself. While exploring the implications of this curious convergence is beyond the immediate aim of this book, perhaps the environment, and impact assessments in particular, might be a site to reconsider the relentlessly normalizing tendencies of Foucault's theory of biopolitics (despite at least one interlocutor's suggestion that Foucault's biopolitics had nothing to do with the environment [Lemke 2011]).

44. Chris Sellers (2012: 9) writes that environmentalism "first appeared in the United States." Even scholars emphasizing the global aspects of environmentalism note the "US intellectual hegemony over the environmental movements as a whole, at least since the 1970s" (Martinez-Alier 2002: 7). For some scholars, this fact is derived from the "post-materialism" thesis. From Marxist historian Eric Hobsbawm to political scientist Ronald Inglehart, this scholarly orientation explains the rise of the environment in America as a symptom of how generalized affluence lifts society out of the cruelties of struggles over food, shelter, and wages and into more privileged aspirations. The only people who care about clean air and clean water are those who are no longer struggling to survive. The environment is evidence not of material problems but of post-material privilege.

Alongside the planetary dimensions of an "environmentalism of the poor," this book counters such tendencies. It was not that the United States was ensconced in an affluence that escaped the gravity of material destitution. Rather, it is precisely the material conditions of that synthetic prosperity that generated new vulnerabilities. Environmentalism began in the United States not for the complete absence of material considerations but for the catastrophic materiality that underwrote the advertised ease of American prosperity. This "Trevelyan thesis," as Martinez-Alier (2002: 9) explains, argues the "appreciation of Nature grew proportionately to the destruction of landscapes wrought by economic growth." As one commentator noted of the rise of the environmental crisis in the 1960s, America possessed a "paradoxical ability" to "devastate the natural world and at the same time mourn its passing" (Ekirch 1963: 189), a fact not unrelated to the US monopolization of industrial production after World War II (when other manufacturing centers in Europe and Japan found themselves bombed into rubble; see Mahrane et al. 2012). "The precedence of the United States in environmental science and policy," writes Joachim Radkau (2014: 91), stems in part "from the fact that the exploitation of resources had long been particularly ruthless there."

45. *Umwelt* gained intellectual popularity in Martin Heidegger's philosophical reflections of the wild in Nazi Germany. Heidegger attributed the concept of *Umwelt* to Wilhelm Dilthey. But as Dieter Thomä (2001: 116) notes, Dilthey was also borrowing the term: "As a matter of fact, Dilthey helps introduce the term *Umwelt* in the German-speaking world: He uses it as a translation of Comte's deterministic concept of *milieu*." The first introduction of *environment* into US social science stands as a rough-hewn amalgam of Dilthey's *Umwelt* and Comte's *milieu*.

For the first half of the twentieth century, "environmentalism" was a distinguished current of thought within US social science, one emphasizing contextual explanation in social research. Gesturing toward *milieu* and *Umwelt*, the term *environmentalism* foregrounded the influence of surroundings in advancing a more scientific understanding of social development. The 1930 edition of the *Encyclopedia of the Social Sciences* has an entry for *environmentalism* (but not *environment*). Here, environmentalism provides a flattened theoretical synthesis of wide-ranging accounts that look to context as the primary coordinates of explanation (grouping together the likes of Charles Darwin, Marx, Franz Boas, and Robert Park). In some ways, the term is a pitch-perfect translation of *milieu* and *Umwelt* into efforts to professionalize social science in the United States, generalizing the insights of context almost to the point of banality. But such generality was, itself, novel. At a moment when university-based research and inquiry is sorting the world into siloed disciplines, environmentalism moved toward a transdisciplinary and trans-species concept, equally at work in investigations of Pavlov's dogs, mutations in fruit flies, the fur pigmentation of Himalayan rabbits, civilizational types in ancient Mesopotamia, and crime statistics in meatpacking Chicago. At times, environmentalism gestures toward a materialist inquiry—"the environmentalism implicit in the writings of Karl Marx"—but it's a gesture that defangs Marxism by excluding histories of conflict and questions of domination. In the end, environmentalism works to generalize an analytic stance toward a field rather neutrally called "context" (in which climatic zones, longitude and latitude, humidity, agriculture, technology, and even culture are lumped). Place comes to reign supreme while historical explanations are neatly avoided. The environment and environmentalism have a more prominent place in the next edition of the *Encyclopedia of the Social Sciences*, which came out in 1967. The *environment* entry summarizes new work in natural ecology (i.e., Odum 1953), while the *environmentalism* entry summarizes the Julian Steward/Carl Sauer cultural ecology project. While the former is still wed to isolated sites in intricate balance (namely, mountain ponds and wetlands), the latter is still tied to the interactive relation of man and nature. Neither of these projects—natural ecology or cultural ecology—analytically foregrounds *environment* or *environmentalism*, and the main point of these terms in the scholarship indexed by these entries is to provide a generic privileging of context before more established scientific questions or politically committed terms break apart curiosity about the world into specific disciplines. Environment is an intellectual sensitivity to context, whether that gravitational pull of context consists of landscape, climate, ecology, or even concepts. While noting that modern man "has become a new geological force" that shapes as much as is shaped by context, environment is taken to be an entirely scholarly optic. The environment does not presume any reality beyond how the scholar views the world through it. For the first half of the twentieth century, "environmentalism" was a rather aloof school of thought within US social science; it was not a substantive problem, a regulatory project, or a stable scientific field. Yet it is interesting to note that many of the leading scholars of this previous iteration of the environment—Julien Steward, Carl Sauer, and others—had seats at the table in those venues where the environment emerged as a new political crisis and scientific jurisdiction in the 1960s (Darling and Milton 1966).

46. Many of the intellectual debates that accompanied the crisis of life provoked by petro-capitalism and thermonuclear statecraft brought together scholars primed in the pedantic version of "environmentalism" and they, in turn, framed many of their insights into the contemporary crisis of life in terms of the environment. And indeed, many of the scholars that participated in the 1956 conference "Man's Role in the Changing Face of the Earth" (Thomas 1956)—a project that took the emerging crisis of life in the present as the perfect perch from which to ask new questions of history—found themselves enlisted a decade later into "Future Environments of North America" (Darling and Milton 1966), a project that asked participants to imagine new policies for surviving the future. Many participants, like Lynton Caldwell, were also drafted into helping write the first environmental legislation. But as the environment gained stately recognition, its intellectual precursor drifted into scholarly irrelevance.

47. Bonneuil and Fressoz (2016) also incisively argue that the presumed novelty of the environment served, like the Anthropocene, to dampen and disavow longer histories of ecological reflexivity in the rise of capitalism.

CHAPTER 2. GOVERNING DISASTER

1. It is worth noting that there is no other federal venue for these concerns to be addressed. The Health and Human Services (HHS) longitudinal health study of the BP oil spill was restricted to emergency workers. Economic aid was focused on those businesses that lost customers due to the spill (and had the receipts to prove it). In both instances, ordinary people sickened in the aftermath of the spill were excluded from official concern. The environment, actively revised during the spill, became an obvious venue where many sickened people came to try to make their ailments intelligible to the force of the state.

2. The environment, like the economy, is remarkably new (Mitchell 1998). Over the course of the past century, the *environment* shifted from an erudite synonym for *surroundings* to a proper noun worthy of its own governing agency in nearly every nation-state (whether that domain is called *environment, medio ambiente, mazingira, lingkungan,* or *huanjing*). While each instantiation has a decidedly local inflection—from forests in India and Romania to water in Bolivia and Uganda to sustainable development in Australia and Argentina to tourism in Namibia and Madagascar to rural affairs in England—each is also put to work rendering disparate biological aspects of specific countries amenable to state governance and commensurable for international forums. This emergent environment, enshrined in the national singular of protective legislation and institutional

practices, then parallels the rise of the national economy as a consequential domain that depicts the vital conditions of each nation-state in a broadly legible manner (a legibility enhanced by its achieved distance from embodied perspectives and historical relations) (Miller and Rose 1990; Mitchell 1998).

3. A number of critical theorists have suggested that the rise of capitalism itself is what renders nature discrete, either as the finite content of the insatiable commodity form or as a serene landscape desired by those whose affluence rests on social unease. The governance of nature, in much of the resulting research, often then centers on "natural resources" like forests or fisheries or even idealized places of nature whose abrupt scarcity has made them a consequential problem (and constitutive power) for the state (Peluso 1992; Watts and Peluso 2001; Robbins et al. 2007). The theoretical dimensions of the commodity are at the forefront of this research, helping bring the distributive logistics and exclusionary logics of natural resource management into clear focus. There is much to be excited about in this research. One point I would like to make, however, is the manner in which such theoretical commitments quickly—perhaps too quickly—interpret the governance of the environment as bringing yet another lively alterity into the sober discipline of the commodity (e.g., West 2006). The emergence of the environment in the United States, I would argue, was less a strategic response to depleted natural resources than it was a series of urgent reactions to industrial disasters (and the knowledge of life's vulnerability they incited). The ultimate effect of the governance of the environment may very well be the uneven distribution of natural resources, health care, public intelligibility, and so forth. But the cause, as I suggest here, following Agrawal (2005), Escobar (2008), and Mathews (2010), should be analyzed as a political formation not immediately reducible to what we already know of the commodity.

4. In 2015 I presented a paper expanding upon this point at the interdisciplinary conference "Thinking the Earth" at Brown University. Speaking off the cuff, I said at one point that for the first month of the BP oil spill federal agencies did not respond to what was happening right in front of them, they responded to the oil spill they already knew how to solve: *Exxon Valdez*. A hand shot up in the back of the room as a middle-aged man in a dark suit rose to speak. I stopped and gave him the floor. "I'm the administrator of Region One in the EPA and I helped lead EPA's efforts along the Gulf Coast during the spill," Curt Spalding said by way of introduction. A pause filled the room, everyone leaning in to hear what he would say next. "And I just want to say, you are spot on." Curt explained there was a binder explaining how to respond to an oil spill. Senior leadership at EPA and NOAA pulled the binder out, opened it up, and did what it said. While this emergency response is highly effective when oil spills mirror the *Exxon Valdez* disaster, it proved immensely counterproductive during the BP oil spill. A few years later, Jane Lubchenco visited Bennington College to give our annual distinguished science lecture. Jane had overseen the entire federal response to the BP oil spill, and a colleague arranged a meeting. I explained the work I had done during the oil spill, whom I spoke with, and the unfolding science I observed close up. I thanked her profusely for the openness with which federal agencies conducted their work, without which my own inquiry and so many others would have been impossible, before

moving into my critique. "The federal government responded to the wrong oil spill," I said, describing how the failures of *Exxon Valdez* overdetermined the official response and blunted the scientific curiosity necessary to see what was actually unfolding in the Gulf during the first few weeks of the spill. "You're absolutely right," Jane said. The federal response to major oil spills is "locked-in place through legislation," and there was little else that could be done until the inadequacy of those protocols were abundant to everyone.

5. The problem with the fluorometer, for Unified Command, was that it had been tinkered with and adjusted to work in an extreme deepwater environment saturated with hydrocarbons. The devices worked, but each worked in a unique way. "Too much emphasis has been put on the Florometer as a prospecting tool," one Unified Command official told me. "In terms of the calibration and validation of instruments, all Florometers are different." "A Florometer produces a signal, not hard data," another official told me. "To be a fact requires uniform sampling and laboratory testing." ("None of our equipment was designed to work in an oil spill," one academic scientist responded.)

6. Subparts per billion, as in 0.25 ppb; anything below that relied more on personal ability to manipulate equipment in the laboratory and could not be easily standardized. Practically, as one official said, "Our current analytical capacity can't track it further than 10–20km" from the wellhead.

7. "BP has committed $500 million over the next 10 years to this thing called science," a BP official noted. Previous funding for marine science in the Gulf of Mexico was less than $10 million annually.

8. What Canguilhem says of disease also, I think, holds true for disasters. "Disease reveals normal functions to us at the precise moment when it deprives us of their existence," Canguilhem ([1966] 1991: 100–101) writes. "Diseases are new ways of life."

9. This relation of disaster to an ex post facto normal is something that American anthropology has some familiarity with. From Lewis Henry Morgan to many students of Boas, the impending doom of indigenous communities shaped the analytical practices of anthropology. (In a 1979 review of the anthropology of disaster, William Torry suggested anthropology's long acquaintance with culture under duress offered practical insight into how to study a disaster.) For that previous generation of anthropology, the pressing task was one of collecting whispered remnants and then carefully reconstructing the autonomous whole that must have existed just before the interventions of empire (and ethnology) arrived. Anthropologists salvaged the baseline of social life as it presumably wasted away. While such an isolated and idealized normal carried dubious epistemological commitments—namely, a principled avoidance of historical entanglements (Wolf 1982; Fabian 1983)—this formatting of normality has proved rather effective for other tasks. The resulting depictions of "normal life" have since found a curious afterlife within many of the communities depicted, not so much as dated portrayals of social life but as fixed definitions of authentic life (Clifford 1988). Such objectifications of normality, analytically framed as separate from the impinging destruction that made them interesting in the first place, have become a rather handy guide to carving out exceptional spaces of political becoming in the present (Bessire 2014). We are only just beginning to realize the rippling consequence

of those urgent measurements of normality instigated by disaster; that is, how an emergency baseline can later come to operate as a persuasive subject position (Cepek 2012) and a potent modality of governance (Fassin 2012).

CHAPTER 3. ETHICAL OIL

1. I spent the month of July in the tar sands in both 2013 and 2014. Based out of Fort McMurray, Alberta, I worked to better understand how the oil companies manage the ecological disasters of their extractive operations. Each of the six tar sand operators I interacted with was immensely proud of the technical infrastructures they had designed to contain those disasters. While these infrastructures of containment were obscenely insufficient for the task at hand—the disastrous reach of extraction exceeded their ability on nearly every front—I was struck by how ideologically reflexive these infrastructures of containment were. They anticipated a preeminent critique of oil emerging from the moral authority of alterity. I interviewed a handful of company officials charged with environmental remediation at three different tar sands companies, took formal tours of five restoration sites, read through hundreds of impact assessments in the local library, and spoke with a few concerned residents and organizations in the region. While my focus was primarily on how the corporations managed the negative ecologies of their operations, I also made short visits to two First Nation villages, Fort McKay and Fort Chipeywan, where I introduced myself to and learned about the immense challenges they face. Yet the main thrust of this chapter centers squarely on how the oil companies themselves take up an ontological theory of cultural ecology in order to both reify and redeem the unbound disaster of their present operations.

2. Operated by three people, a single bucketwheel excavators can move more earth than was required in the digging of the Panama Canal.

3. EROI is industry-speak for "energy returned on investment"—that is, how many units of energy were required to extract units of energy. The standard ratio in the oil industry is 1:25, which is to say the energy of 1 barrel of oil is required to extract 25 barrels of oil from the earth. According to tar sands operators, that ratio falls to 1:5 due to the immense amount of natural gas used to separate the sand from the bitumen (indeed, many of the pipeline projects underway when I visited were to bring more natural gas to the tar sands). Others calculate the EROI of the tar sands at closer to 1:2.9, or even—if we include the energy required for distribution and remediation of the extraction sites—closer to 1:1 or lower (Nuwer 2013: 1)

4. The 1899 negotiations for Treaty 8 pivoted on these subsistence rights, as leaders from the Dene and Cree of Fort McKay, Fort Chipywan, and Fort McMurray made clear: "Complete freedom to fish. Complete freedom to hunt. Complete freedom to trap" (cited in Fumoleau 2004: 77).

5. Whether from irony or ignorance, a number of the cultural impact statements of tar sands projects have come to rest their analytical definition of *tradition* on Eric Hobsbawm's essay "Inventing Traditions."

6. Timoney and Lee (2011: A) write: "The view of the Alberta Government, the primary agency responsible for environmental management in the region, is

that industrial activity is not detectable in the concentrations of PAHs and other contaminants observed in the region." One journalist summarized the government's position this way: "Both Ottawa and the Alberta government, which are dependent on oil sand taxes and royalties, claim that all the pollution in the river is naturally occurring based on proprietary data collected by an industry-funded group" (Nikiforuk 2010: 1).

7. "This project emerges from a collaboration initiated by the Mikisew Cree First Nation and Athabasca Chipewyan First Nation in northern Alberta and with scientists from University of Manitoba and University of Saskatchewan" (McLachlan 2014: 10).

8. A handful of Native corporations operating in the tar sands now bring in over $1 billion annually in services rendered. Many First Nation corporations specialize in housing workers and disposing of chemical waste, two tasks that executives at oil companies have indicated they were happy to shed. Noting that hunting and fishing in the region have become impossible, Dave Tuccaro, CEO of one of the largest First Nation corporations, argues that the tar sands "have to be our new livelihood, our new trap lines" (Vanderklippe 2012: 1) Other communities are a bit more hesitant. The president of Primco, a Dene Corporation working in the tar sands, explained his community's decision to partner with oil companies. James Blackman noted, "There's nothing healthy about it at all. . . . Industry pushes through regardless. We have to work with them collectively to try to at least get a better livelihood for the loss of the land" (quoted in Vanderklippe 2012: 1).

9. This point was echoed in interviews with officials from other oil companies. One told me: "There is significant natural seepage along the Athabasca River. Who knows what that stuff is doing. Our operations are actually cleaning up nature by removing that dirty stuff from the landscape." In 2014, a team of environmental scientists devised an analytical method to distinguish petrochemical contamination originating in natural seeps from petrochemical contamination originating in mining operations and tailing ponds. The results found mining operations were a substantial source of petrochemical contamination of the Athabasca River, each day flushing millions of gallons of carcinogen-laden wastewater into the region's waterways (Frank et al. 2014).

10. It is worth noting that against the nuanced and often quite diverse terrain of Native identity in this region, for the companies "Indigeneity" has been standardized into a single vision of the primordial boreal forest.

11. These findings were no surprise for many restoration scientists. E. A. Johnson had noted in 2008 that the scale of restoration in the tar sands greatly exceeded any previous project. Far from restoring a modest drilling pad or pipeline path back to the landscape, the tar sands called for the restoration of entire landscapes. Johnson told one reporter (Why Files 2008): "Restorations are usually small projects, a few hectares in size, but now we are confronted with whole landscapes in which the reconstruction must start with the central template, the groundwater, and then the soil. . . . We are going to have to reconstruct the drainage, the ground-water flow, and these are things about which we have little knowledge. It is not clear to me that everybody understands how complicated this is" The

devastation being enacted by extraction calls for a scale of restoration previously untried: terraforming (Johnson and Miyanishi 2008).

12. Wadsworth wrote up his results in an internal report, which was soon leaked to several newspapers. The AER quickly issued a retraction, saying Wadsworth's calculation was based on "a hypothetical worst-case scenario" and the actual liability was a much more manageable sum well within the ability of oil companies to fund from their projected profits (Bellefontaine 2018: 1).

CHAPTER 5. PETROCHEMICAL FALLOUT

1. After the discovery of PFOA, many people arrived on the scene with full knowledge of how bad it was. Since at least *Silent Spring* (if not the Book of Genesis), the discovery of toxicity has been a story we know how to tell. And the genre itself has become a powerful force: an idyllic community suddenly waking up to the horrors of its own destruction. Many of the national journalists and environmental activists who visited the community in the aftermath of PFOA arrived and departed with this story. And they often did so to great fanfare. From the *New York Times* to Erin Brockovich, PFOA was a story of spectacular harm and innocence lost. PFOA, one journalist told me, is "a petrochemical Fukushima." So often, it's this retrospective understanding that frames popular and scholarly understandings of toxicity. But this was not the story the impacted community told themselves. For many residents, PFOA was not a spectacular event so much as long-standing fears slowly given the feeling of fact. And I was struck by how residents quietly resisted the given genre even as they found themselves drawn to its scripted perspective when national media and advocacy groups visited.

2. 3M was "the primary if not sole manufacture[r]" of PFOA, as subsequent lawsuits have put it (Grandjean 2017: 5), which it produced at its Cottage Grove plant outside of Minneapolis from 1947 until 2002. During that time, the largest purchaser of PFOA was DuPont, which used PFOA in the production of Teflon products at its Washington Works plant in Parkersburg, West Virginia, from 1951 until the EPA facilitated a phaseout of PFOA in 2015. Throughout this history, both 3M and DuPont disposed of PFOA in unlined pits on their respective plant properties, in nearby municipal or private dumps, by shipping it out for unregulated disposal at sea, and by emitting it by the ton from stacks at the plant.

3. In the 1960s, a pharmacologist studying inorganic fluorine in human blood found growing evidence of organic fluorine in his samples, for which there was no known natural source. Donald Taves published his results in *Nature* in 1968 under the title "Evidence That There Are Two Forms of Fluoride in Human Serum," noting that this new "fluorocarbon molecule [. . .] does not break down or diffuse [. . .] implying a large stable molecule" (1051). Nearly ten years later, Taves and a colleague were comparing fluorine blood levels of people living in Texas (with high levels of inorganic fluorine in the water) and people living in New York (with low levels of inorganic fluorine in the water). Comparing the samples, the scientists discovered both populations had detectable

and strikingly similar levels of "fluorocarbon carboxylic acids" in their serum, which they both deduced were of synthetic origin. The professors called 3M in 1975 and asked "where such a 'universal' presence of such compounds could come from," noting such compounds were not present in laboratory animals, and perhaps consumer goods like "Teflon cookware" or "Scotchguard fabrics" could be the source. The question caused quite a stir at 3M, and an internal 3M memo notes that company officials "pleaded ignorance," falsely advised the scientists that their products contained no organic fluorine, and warned them "not to speculate" about the source of fluorocarbons (Lerner 2018).

4. Wilbur Tennant made home videos of his cows. These videos have been described by reporters: "Another blast of static is followed by a close-up of a dead black calf lying in the snow, its eye a brilliant, chemical blue. 'One hundred fifty-three of these animals I've lost on this farm,' Wilbur says later in the video. 'Every veterinarian that I've called in Parkersburg, they will not return my phone calls or they don't want to get involved. Since they don't want to get involved, I'll have to dissect this thing myself. . . . I'm going to start at this head.' The video cuts to a calf's bisected head. Close-ups follow of the calf's blackened teeth ('They say that's due to high concentrations of fluoride in the water that they drink'), its liver, heart, stomachs, kidneys and gall bladder. Each organ is sliced open, and Wilbur points out unusual discolorations—some dark, some green—and textures. 'I don't even like the looks of them,' he says. 'It don't look like anything I've been into before.' Cows with stringy tails, malformed hooves, giant lesions protruding from their hides and red, receded eyes; cows suffering constant diarrhea, slobbering white slime the consistency of toothpaste, staggering bowlegged like drunks. Tennant always zoomed in on his cows' eyes. 'This cow's done a lot of suffering,' he would say, as a blinking eye filled the screen" (Rich 2016: 25).

5. The voluntary nature of this phaseout allowed DuPont to continue contaminating nearby drinking water sources without penalty or responsibility. As Sharon Lerner (2015) has written: "Without a legally binding agreement, the EPA was unable to enforce the terms of its interactions with DuPont." While DuPont agreed to provide clean drinking water to residents whose water contained dangerous levels of PFOA, the company kept moving that threshold of concern upward. DuPont's internal standard for drinking water was 1 ppb. It initially agreed to provide drinking water to those above 14 ppb. "Two months later, that 'trigger level' was raised to 150 ppb, where it stayed until 2006" (Lerner 2015).

6. A notoriously cranky machine prone to constant tinkering, the mass spectrometer works like this: a sample is injected into a carefully orchestrated cyclone of electrical charges that drive one exacting atomic weight like a bullet through the machine while casting all other weights aside like debris in a tornado. At the end of the machine, the surviving elements slam into a delicate plate that emits a precise cascade of electrons in relation to the atomic weight of the substance impacting it. These pulses are then translated into measures of the atomic weight. It has been described, not unfaithfully, as "the blunderbuss" of laboratory instrumentation, using blunt force to choreograph a dance of atoms. Mass spectrometry first took consequential shape through its ability to weigh the minute differences in mass among various uranium isotopes,

becoming instrumental to the purification of weaponizable isotopes of uranium in the Manhattan Project. As John Fenn (quoted in Siuzdak 2004: 51), who was awarded a Nobel Prize for his work advancing the sensitivity of the device, has said, "Mass spectrometry is the art of measuring atoms and molecules to determine their molecular weight." In the last thirty years and with advances in how a sample can be stabilized through techniques that ionize and spray—the "liquid" in liquid mass spectrometry—a chemical through the device, "the sensitivity and speed of modern day mass spectrometers has increased nearly six orders of magnitude." These advances have pulled mass spectrometry deep into environmental science and policy.

7. Indeed, as PFOA was being phased out of US plastics manufacturing in 2015, most firms simply tinkered with the recipe and incorporated a similar, but molecularly distinct, PFAS compound into their operations. Advocates and state agencies have requested LC-MS standards for what the Environmental Working Group calls "artisan PFAS compounds." Companies have been uniform in their responses: those customized PFAS compounds are proprietary and cannot be shared. (This is not a problem for PFAS alone; there are some eighty thousand unregulated chemicals in circulation. As one leading environmental engineer told me, "The cost of developing LC-MS standards for each and implementing robust testing of sites where they are used would likely exceed the market value of many of these chemicals.")

8. For some time, work in the anthropology of science has found inspiration in the suspicion that disciplined ways of knowing may play an understated role in producing the object of inquiry. From Bruno Latour to Ian Hacking to Lorraine Daston, the formal manner or mechanism of knowing a thing can confer new stability on that thing, a stability that is both previously unknown and now definitive of the thing itself. What, then, are we to do with PFOA, this synthetic chemical whose engineered stability now exceeds the devices and visions of the laboratory—indeed, whose engineered stability and negative ecologies seem to exceed the wildest dreams of objectivity itself? The crucial problem of PFOA may not be the stability conferred by scientific methods but the disastrous durability of the petrochemical itself, which our best means of knowing and acting seem to but dimly grasp. The pressing question, as Adriana Petryna (2015) writes so movingly of the supercharged forest fires of our climatic present and the forest fighters struggling to keep up, may now be: How can we become ethnographically attuned to that uneasy orbit between the unloosed materiality of our contemporary world and our ever-imperfect methods of concern?

9. I played a minor role in this work, and much of the research that forms the basis of this chapter draws from five years of personal and professional involvement in this issue (Bond 2021a). As is now understood, the petrochemical PFOA was emitted for half a century from three plastics factories in Hoosick Falls and Petersburg, New York, and North Bennington, Vermont. While corporate owners suspected the health impacts of PFOA for decades and learned of the elevated levels of PFOA in the groundwater years ago, they kept these alarming facts to themselves. Today, it is estimated that these three modest plastics plants contaminated roughly 250 square miles of soil and groundwater (including my own home and the college campus where I teach) in a rural area where many

still depend on agriculture and get their drinking water from wells. In response to the work of Michael Hickey and other outraged residents in 2015, I joined with colleagues Janet Foley (chemistry) and Tim Schroeder (geology) to figure out what a college can—indeed, what a college should—do in a situation of extensive regional contamination, real public health concerns, confused directions from state agencies, and well-heeled corporate subterfuge. With the support of a National Science Foundation RAPID Response Grant, we organized a handful of activities that put the analytic resources of the college to work responding to community concerns. This included the following:

(1) *Understanding PFOA*: We designed a primer class on the chemical composition, environmental pathways, health concerns, and regulatory status of the PFOA . We have offered this class every spring to local residents free of charge.

(2) *Free Analysis of Drinking Water*: Many folks rely on residential wells in this region. Initially, the state would only test wells in the immediate vicinity of the plastics factories. We offered residents free analysis of their drinking water (and soon identified illegal dumps of PFOA and far more extensive contamination than state models initially allowed for).

(3) *Community Health Questionnaire*: Exposure to trace amounts of PFOA is strongly linked to testicular cancer, kidney cancer, pancreatic cancer, and thyroid cancer (among others). In 2017, NYS DOH did a desktop search of the state cancer registry and found no elevated levels of related cancers in Hoosick Falls. Yet community members knew of numerous local residents afflicted with these cancers. Collaborating with doctors and public health experts, we organized a community health questionnaire to give the community's knowledge of its own health more prominence in ongoing conversations.

(4) *Confronting Corporate Antics*: Saint-Gobain, one of the offending corporations, has invested heavily in various devious strategies of intimidating residents and spreading blame for the contamination onto local communities. We've tried to figure out how we might both draw these investments of the corporation into a more public light and conduct environmental research that might better attribute the immense scale of contamination to specific industrial sources.

(5) *Op-eds*: In response to the technical complexity of the official response to PFOA contamination (and the manner in which one legal strategy of polluting industries seems to be amplifying this complexity), we've authored a handful of op-eds in local papers that work to summarize the significance of various reports, share scientific research that troubles some of the corporate arguments, and advocate for better health resources for impacted communities.

As I've turned from four years of advocacy on this issue to write about it, I've been struck by how prominent toxicity is becoming in certain currents of anthropological theory *and* how little those theories seemed to illuminate the lived contamination I witnessed and continue to participate in. This chapter reflects on this disjuncture.

10. This was not as easy as it may sound. At the time, most commercial laboratories did not accept samples for PFOA analysis from citizens as a matter of policy. Too often, as it was explained to me, working with citizens meant having to explain every single thing, a civic labor no commercial laboratory was interested in. Hickey was relentless and pursued several labs until a Canadian firm eventually agreed to run his samples.

11. Later, I wrote about this "uncertainty" in an op-ed. A senior EPA official under President Barack Obama reviewed a draft and told me to avoid the term: "It sounds like we don't know what's wrong with PFOA." I explained my reasoning. "I don't like it. Cut it." The tactical pursuit of justice had no room for these initial experiences.

12. The defining quality of scholarship on "new materialism," in one programmatic statement (Coole and Frost 2010: 8), "is its antipathy towards oppositional ways of thinking." Yet in overcoming modern dualisms, such work can nonetheless inscribe a more founding dualism onto the world: again and again, such work presumes everything can be reduced to either (ideological) purity or (empirical) hybridity. As Elizabeth Roberts (2017) writes, the choice to stay with the trouble has always been a privileged one, and for many marginalized people some borders still offer modest refuge from an existence otherwise under imperial forms of surveillance and control. Andreas Malm scrutinizes hybridity from a different angle. In a moment of profound and proliferating inequity, Malm asks: Should the task of progressive scholarship really be so bound up with dissolving contrasts? Today "one cannot afford *not* to draw lines of separation," writes Malm (2018: 189), bemoaning how hybridity so often ignores the oppositional analytics and politics of class in a moment of profound inequality.

13. This is not a pointed critique of queer theory nor a call for its wholesale rejection. Jain (2007) demonstrates how instrumental queer theory can be in advancing an ethnography of toxicity without losing a sense of outrage over its profitable complicities. Indeed, queer theory has been instrumental in displacing the structural binaries that tether the world to tyrannical hierarchies and in cultivating a non-normative ethics for anthropology (Boellstorff 2007; Dave 2012). As Margot Weiss (2016) argues, queer is a critical theory and disposition, not a fixed identity and even less an institutional project to consolidate power. The point being made in this chapter is how toxics are catapulting one branch of queer theory into a kind of liberation toxicity, where the world must be made queer over and above the avenues of informed consent, democratic politics, and even historical revolution.

14. Pollock's argument about the emancipatory promise of toxicity forms a core part of the programmatic call for "Chemo-Ethnography" in a special issue of *Cultural Anthropology* in 2017, edited by Nicolas Shapiro and Eben Kirskey.

15. Carrying echoes of Kim Fortun's (2014 :312) incisive point: "The complicated fact that even if we have never really been modern, we still have a modernist mess on our hands."

CHAPTER 6. THE ECOLOGICAL MANGROVE

1. This chapter draws on four months of ethnographic and archival research conducted in St. Croix in 2011 on the impacts of the Hess Refinery, and two weeks of archival research in San Juan, Puerto Rico, on the *Commonwealth of Puerto Rico v. SS Zoe Colocotroni* court case, also in 2011. This research provoked new questions about refineries and mangroves that I worked to answer in extensive reviews of journalistic accounts of the building of the Hess Refinery and scientific publications on mangroves in the Caribbean.

2. Excluding the militarized burning of the Kuwait oil fields during the First Gulf War, the four largest oil spills in human history have occurred in the wider Caribbean region. Dos Bocas on the Mexican coast spilled approximately 420 million gallons in 1938; the BP oil spill in the Gulf of Mexico released about 175 million gallons in 2010; Ixtoc I off the coast of Mexico spilled some 145 million gallons in 1979; and the *Atlantic Empress I* off Trinidad's coast spilled approximately 88 million gallons in 1979. By way of comparison, the 1989 *Exxon Valdez* spill—widely seen in the United States as a worst-case scenario oil spill—released a relatively modest 11 million gallons of crude oil.

3. For a parallel argument about mangroves and climate change, see Vaughn (2017).

4. There are important distinctions to be made in the critique of the plantation. Caribbean scholarship has long presumed that its creative edge lay in exposing the deep commensurability between plantation pasts and the modern present. Many new imperial alignments have gained local justification in the wedge they drive between plantation pasts and modern futures.

5. The Oil Workers Strike of 1945—an event a federal official described as "the first nation-wide strike in the history of the oil industry" (Hoch 1948: 117)—and the formation of the two-hundred-thousand-member Oil, Chemical, and Atomic Workers International Union in 1955, demonstrated that the nation's petroleum infrastructure was not intrinsically immune to the demands of labor. Refineries became an opportune place to strike. However, as Matthew Huber (2013: 62) has shown, the resulting upsurge in walkouts at refineries during the 1950s and 1960s gifted oil companies with an ironic insight: "Refinery Strikes Suggest Plants Can Be Run with Still Fewer Men," ran the title of a 1962 *Oil and Gas Journal* article. In the postwar period a new point of scrutiny for domestic refineries also emerged: water and air pollution. As expanding suburbs came to surround refineries, many cities and states began to turn a more critical eye toward the environmental and public health impacts of refineries (Gorman 2001).

6. It is worth recalling that until the recent hydro-fracking boom, the last newly constructed refinery in the United States was built in the 1950s, with the noted exception of sizeable expansions of a few existing port refineries in Louisiana and Texas in 1976 to better accommodate imports.

7. Sidney Chernick, the World Bank's chief of mission to the Caribbean, authored a plan for regional development that emphasized the key role of "enclave-type" processing of petroleum products for export. Such refineries and petrochemical plants would enable a regional shift away from agriculture and begin formatting Caribbean society for the modern economy. While Chernick (1978: 139) acknowledged that "little value" might actually accrue locally, he argued that the most lucrative payoff might be the social discipline such industry imposed upon Caribbean societies. "To encourage enclave exports is not inconsistent with taking longer-term steps to transform the structure of the economy by developing stronger internal behavior."

8. If racialized labor no longer provided the locus of power around oil, the empire of oil certainly did not do away with race. Rather, it introduced a new rubric of racial inequality: unequal exposure.

9. Under colonial law at the time, the US president appointed the governor of the US Virgin Islands, while the legislature was elected locally. The Kennedy family was reputed to have deep ties with the Paiewonsky family of St. Thomas in the (illicit) transport of Caribbean rum during Prohibition.

10. The infamous collapse of sugar prices in the 1920s led to a severe depression in the Caribbean. In the US colonies, this depression became an opportunity to experiment with a state (re)organized economy. US Caribbean colonies became a laboratory for later New Deal interventions on the mainland (Tyson 1991). On St. Croix, the colonial government seized large (and largely abandoned) sugar estates and erected a modern central sugar factory. Land was redistributed into "homesteads": five- to ten-acre plots for largely Afro-Caribbean families to raise sugar cane for export and enough other crops to subsist on. Mechanized farming equipment like tractors and trucks was shared among the homesteaders, and families were promised title to the land. These farms flourished. By 1950, such small-scale farming loosely organized around the central sugar factory was a dominant form of employment and production on the island. Beginning in 1965, the colonial government started turning over the land and equipment to Harvey Aluminum and Hess Oil. The modern sugar factory was dismantled and shipped to Venezuela (where it is said to still operate). The collectively owned farming equipment, purchased by Harvey for exactly ten dollars, was driven into the sea (today the rusted-out remnants are a marginal diving attraction on the island). Before the sale, several communities of farmers were promised that they could remain on the land. After the sale, they were told to leave. Many chose to stay. "They ripped the roofs off the houses and left everything to rot," one farmer recalled in an interview. Although the local newspapers did not take note of this egregious episode, they did report on a strangely orchestrated arson spree that burned sugar cane and other crops to the ground just days before what promised to be a record harvest (*Daily News* 1965b). While Governor Paiewonsky (1990: 340) solemnly announced "the death of agriculture" on St. Croix, a more apt verdict might be that agriculture was murdered (Bond 2021b).

11. This quote comes from a *Virgin Island Times* banner article (1964a). When local newspapers reported the widespread protests against the industrial takeover of St. Croix's farmland, Harvey Aluminum responded by commissioning its own newspaper. The *Times* served as a cloaked mouthpiece for the industry's interests, and though its editor was a Harvey vice president and its entire staff were plant employees, it revealed nothing about its corporate affiliation. The paper promised "to be a happy newspaper, pointing out some of the brighter things that make life worth living. It will avoid a heavy diet of doomsday philosophy" (*Virgin Island Times* 1963: 4). With headlines like "Harvey Big Company?" and "VI Corp Is Fast Fading Away" (VI Corp was the central sugar factory and farming co-op), the paper celebrated industrial development and chastised any local politician who questioned the manufactured demise of agriculture. It compared one senator to Hitler and Stalin for organizing a rally of two thousand people in support of agriculture (1964b).

12. This was a popular sentiment among many I interviewed. "No one really ever thought that cane cutters would suddenly be operating heavy machinery,"

one Crucian recalled. But interviews with retired refinery workers who were recruited from other islands told a different story. "I had no skills when I was hired," a retired Trinidadian worker explained, describing the extensive technical training he and other foreign workers received upon arrival on St. Croix in the 1960s. (They were also trained in the ease with which they could be deported if they caused any trouble.)

13. "Bonded aliens," as imported workers were classified, were housed next to the refinery in camps surrounded by barbed wire fences. These "bonded aliens" could not vote and were denied access to schools and other public services, and their employer could deport them at will. While these workers were initially brought in to serve the island's seasonal tourism trade, the refinery quickly took advantage of this depoliticized class of worker. In 1968, as the refinery underwent a massive expansion, "bonded aliens" accounted for nearly half of the private sector workforce on St. Croix ("Bonded Aliens" 1968: 42).

14. Between 1950 and 1970, refineries in Curaçao doubled their output of petroleum products, while employment in the sector fell from eleven to three thousand. After Puerto Rico's refining and petrochemical boom in the 1960s, unemployment shot up to 23 percent and would have been much higher if migration to the mainland had not provided a crucial outlet for unemployed workers (Powell 1973).

15. *Commonwealth of Puerto Rico v. SS Zoe Colocotroni*, 456 F. Supp. 1327 (1978).

16. *Commonwealth of Puerto Rico v. SS Zoe Colocotroni*, 628 F.2d 652 (1980).

17. While some described this rising appreciation for mangroves as environmental enlightenment spreading from First World to Third World, from metropole to colony— "The wave of environmental concern that began in the temperate regions is spreading to the tropics" (Johannes and Betzer 1975: 1)— the actual geography of insight is a good deal more muddled. While certain aspects of environmentalism, like concerns over clean air and water, do seem to have emerged from industrial cities in North America and Europe, others aspects, like conservation (Grove 1995), climate science (Masco 2010), and as I argue here, interest in the ecology of mangroves, have deep roots in the experimental work of empire.

18. NOAA's Environmental Sensitivity Indices, a popular environmental rubric used to prepare for and respond to oil spills around the world, list mangrove forests as the tropical habitats most sensitive to hydrocarbon pollution (Hoff 2014).

19. Oil spills were by no means the only venue that sparked this new scientific regard for mangroves. However, many of the other sites that formulated a newfound ecological appreciation of mangroves also had deep connections to their wanton destruction. This disastrous epistemology of the mangrove also has roots in Vietnam, where Agent Orange devastated 250,000 acres of tidal forests during the war; in Brazil and Singapore, where exploding coastal cities expanded on land "reclaimed" from mangroves; in Bangladesh, where many have suggested the devastation of the 1970 cyclone (which killed an estimated

450,000 people) would have been greatly moderated had coastal mangroves been left standing; and in countless other tropical estuaries where mangral habitats were razed and cordoned off into private shrimp farms. While oil spills in the Caribbean were far from the only disaster by which the ecology of mangroves became factual and operable, they formed one of the key laboratories for scientifically documenting the worth of mangroves.

20. *Commonwealth of Puerto Rico v. SS Zoe Colocotroni*, 628 F.2d 652 (1980).

21. *Commonwealth of Puerto Rico v. SS Zoe Colocotroni*, 628 F.2d 652 (1980).

22. *Commonwealth of Puerto Rico v. SS Zoe Colocotroni*, 628 F.2d 652 (1980).

23. The energy needs of Caribbean islands that accommodated enclave refineries were underwritten by the glut of crude oil passing through. In this, the Caribbean became one of the few regions in the world (alongside the Middle East) to generate the majority of its electricity from burning petroleum. As Caribbean refineries have begun closing, these energy infrastructures are proving to be quite inflexible to new realities. Nearly 100 percent of St. Croix's electricity now comes from petroleum-fired power plants (Energy Information Administration 2016a). In Puerto Rico, 51 percent of electricity is generated by burning oil, in Haiti 80 percent, and in Jamaica 92 percent (in the United States it is 0.8 percent, and the worldwide average is closer to 5 percent) (Energy Information Administration 2016b; World Bank 2017). This is why the Caribbean pays some of the highest rates for electricity in the world.

24. As refinery workers recalled in interviews, the refinery had a frightful history of environmental disregard. Benzene and other carcinogenic hydrocarbons were occasionally vented out without flaring, and dangerous emissions were routinely released under the cover of storms or night. This was on an island where many residents still get their drinking water from rain catchments and cisterns. Until Occupational Safety and Health Administration posters displaying the dangers of chemicals to workers were put up in the 1980s, mercury was routinely flushed down a drain that emptied into a nearby lagoon near a popular local fishing spot. While fires were a recurring problem at the refinery— the refinery had long housed its own firefighting squad, in part to specialize in refinery fires and in part to keep problems off the public radar—in 2011 a series of huge explosions rattled the island and shut down nearby schools. These explosions drew new scrutiny from the EPA, which uncovered a disconcerting history of toxic releases and environmental shortcuts. Facing potentially record-breaking fines, HOVENSA settled with the agency. The refinery agreed to pay a $5.3 million fine ($5.1 million of which went to the federal government) and, in lieu of penalties for its extensive history of contamination on the island, negotiated a settlement committing the plant to spend $700 million in capital improvements targeted at efficiency updates and environmental protections (EPA 2011). About ten days after the agreement was signed, the refinery announced it was closing and thereby avoided paying for capital improvements and environmental protections.

CONCLUSION: NEGATIVE ECOLOGIES AND THE DISCOVERY OF THE ENVIRONMENT

1. Surely the upsurge of enchantment in theory and practice has something to do with the dim realization that our modern reign has not so much expelled the spirits from our disenchanted reality as conjured up a new regime of haunts, synthetic ghosts entirely of our own making but just beyond our ability to command or even ally ourselves with. Alongside nuclear weapons, fossil fuels and petrochemicals have been the primary instigators of this toxically enchanted world. Yet such dismal realities are more than fodder for theory; they are also lived.

References

Achakulwisut, Pattanun, Michael Brauer, Perry Hystad, and Susan Anenberg. 2019. "Global, National, and Urban Burdens of Paediatric Asthma Incidence Attributable to Ambient NO2 Pollution: Estimates from Global Datasets." *Lancet Planetary Health* 3(4): 166–78.

Adorno, Theodor. (1966) 2007. *Negative Dialectics*. Translated by E. B. Ashton. New York: Continuum.

Agard-Jones, Vanessa. 2016. *Cultures of Energy*. Podcast, episode 35, September. https://cenhs.libsyn.com/2016/09.

Agrawal, Arun. 2005. *Environmentality: Technologies of Government and the Making of Subject*. Durham, NC: Duke University Press.

Ah-King, Malin, and Eve Hayward. 2013. "Toxic Sexes: Perverting Pollution and Queering Hormone Disruption." *O-zone: A Journal of Object-Oriented Studies* 1:1–12.

Ahmann, Chloe. 2018. "'It's Exhausting to Create an Event Out of Nothing': Slow Violence and the Manipulation of Time." *Cultural Anthropology* 33(1): 142–71.

Alberta Government. 2010. "Tell It Like It Is!" Oil Sands Fact Sheet. Edmonton, AB.

Allen, Barbara. 2003. *Uneasy Alchemy: Citizens and Experts in Louisiana's Chemical Corridor Disputes*. Cambridge, MA: MIT Press.

Altman, Rebecca. 2022. "How Bad Are Plastics, Really?" *Atlantic*, January 3, 23–29.

Appel, Hannah. 2012. "Offshore Work: Oil, Modularity, and the How of Capitalism in Equatorial Guinea." *American Ethnologist* 39(4): 692–709.

———. 2019. *The Licit Life of Capitalism: US Oil in Equatorial Guinea*. Durham, NC: Duke University Press.

Appel, Hannah, Arthur Mason, and Michael Watts, eds. 2015. *Subterranean Estates: Life Worlds of Oil and Gas*. Ithaca, NY: Cornell University Press.

Apter, Andrew. 2005. *The Pan-African Nation: Oil and the Spectacle of Culture in Nigeria*. Chicago: University of Chicago Press.

Asimov, Isaac. 1970. "Environment: Have We Committed the Irrevocable?" *Boston Globe*, September 13.

Asper, Vernon. 2010. "Vernon Asper's Daily Reports from 'Ground Zero' of the Gulf Oil Spill." May 12. https://olemiss.edu/depts/mmri/programs/rapid.html.

Atwood, D. K., F. J. Burton, J. E. Corredor, G. R. Harvey, A. J. Mata-Jimenez, A. Vasquez-Botello, and B. A. Wade. 1987. "Results of the CARIPOL Petroleum Pollution Monitoring Project in the Wider Caribbean." *Marine Pollution Bulletin* 18(10): 540–48.

Austin, D. Andrew. 2020. "Economic and Fiscal Conditions in the U.S. Virgin Islands." Congressional Research Service Report, February 13.

Auyero, Javier, and Debora Alejandra Swistun. 2009. *Flammable: Environmental Suffering in an Argentine Shantytown*. Oxford: Oxford University Press.

Baker, Janelle, and Clinton Westman. 2018. "Extracting Knowledge: Social Science, Environmental Impact Assessment, and Indigenous Consultation in the Oil Sands of Alberta." *Extractive Industries and Society* 5(1): 144–53.

Balée, William. 1994. *Footprints in the Forest: Ka'apor Ethnobotany—The Historical Ecology of Plant Utilization by an Amazonian People*. New York: Columbia University Press.

Barry, Andrew. 2013. *Material Politics: Disputes along the Pipeline*. Chichester, UK: Wiley-Blackwell.

Barry, Tom, Beth Wood, and Deb Preusch. 1984. *The Other Side of Paradise: Foreign Control in the Caribbean*. New York: Grove Press.

Bataille, Georges. (1967) 1991. *The Accursed Share*. Cambridge, MA: Zone Books.

Beck, Ulrich. 1993. *Risk Society*. London: Sage Publications.

Beckert, Sven. 2015. *Empire of Cotton: A Global History*. New York: Vintage.

Beckford, George. 1972. *Persistent Poverty: Underdevelopment in Plantation Economies of the Third World*. Kingston, Jamaica: West Indies University Press.

Belcher, Oliver, Patrick Bigger, Ben Neimark, and Cara Kennelly. 2019. "Hidden Carbon Costs of the 'Everywhere War': Logistics, Geopolitical Ecology, and the Carbon Boot-Print of the US Military." *Transactions of the Royal Geographical Society* 45(June): 65–80.

Bellefontaine, Michelle. 2018. "$260B Liability Figure for Abandoned Energy Infrastructure an 'Error in Judgement': AER." CBC News/Radio Canada, November 1. https://www.cbc.ca/news/canada/edmonton/alberta-energy-regulator-liability-figure-error-1.4888532.

Benjamin, Walter. (1940) 2002. *The Arcades Project*. Translated by Howard Eiland. Cambridge, MA: Harvard University Press.

Bennett, Jane. 2009. *Vibrant Matter: A Political Ecology of Things*. Durham, NC: Duke University Press.

Benson, Peter, and Stuart Kirsch. 2010. "Capitalism and the Politics of Resignation." *Current Anthropology* 51(4): 459–86.

Bernabé, Jean, Patrick Chamoiseau, and Raphaël Confiant. 1989. *Éloge de la Créolité*. Paris: Gallimard.

Bessire, Lucas. 2014. *Behold the Black Caiman: A Chronicle of Ayoreo Life*. Chicago: Chicago University Press.

———. 2021. *Running Out*. Princeton, NJ: Princeton University Press.

Bessire, Lucas, and David Bond. 2014. "Ontological Anthropology and the Deferral of Critique." *American Ethnologist* 41(3): 440–56.

Best, Lloyd. 1968. Outlines of a Model of Pure Plantation Economy. *Social and Economic Studies* 17(3): 283–324.

Best, Lloyd, and Kari Levitt. 2009. *The Theory of Plantation Economy: A Historical and Institutional Approach to Caribbean Economic Development*. Kingston, Jamaica: Ian Randle.

Bienaimé, Pierre. 2016. "Here's How Many 'Super Nukes' American Scientists Thought It Would Take to End the World in 1945." *Business Insider*, December 11, 1–3.

Birkeland, Charles, Amanda Reimer, and Joye Young. 1976. *Survey of Marine Communities in Panama and Experiments with Oil*. Washington, DC: Environmental Protection Agency.

Blake, Mariah. 2015. "Welcome to the Beautiful Parkersburg, West Virginia." *Huntington Post Highline*, August 27, https://highline.huffingtonpost.com/articles/en/welcome-to-beautiful-parkersburg/.

Boellstorff, Tom. 2007. "Queer Studies in the House of Anthropology." *Annual Review of Anthropology* 36(October): 17–35.

Boltanski, Luc, and Eve Chiapello. 2007. *The New Spirit of Capitalism*. Translated by Gregory Elliott. London: Verso.

Bond, David. 2013. "The Environment Must Be Defended: Hydrocarbon Disasters and the Governance of Life during the BP Oil Spill." PhD diss., Department of Anthropology, New School for Social Research.

———. 2021a. "Understanding PFOA." *Medical Anthropology Quarterly*, Critical Care Series (November): 1–3.

———. 2021b. "St. Croix at the Crucible: Histories of an Impossible Present." *St. Croix Source*, May–June. https://www.bennington.edu/news-and-features/climate-justice-begins-places-saint-croix.

———. 2021c. "What's Wrong with the White Working Class." *Anthropology Now* 13(1): 37–43.

Bond, David, and Jorja Rose. 2018. "Saint-Gobain's Claims Don't Hold Water." *Vermont Digger*, May 20.

"Bonded Aliens Make up Nearly Half the Work Force in the U.S. Virgin Islands." 1968. *Monthly Labor Review* 91(12): 40–41.

Bonneuil, Christophe, and Jean-Baptiste Fressoz. 2016. *The Shock of the Anthropocene: The Earth, History and Us*. Translated by David Fernbach. London: Verso.

Bookchin, Murray. 1962. *Our Synthetic Environment.* New York: Knopf.

Brosius, J. Peter. 1999. "Analyses and Interventions: Anthropological Engagements with Environmentalism." *Current Anthropology* 40(3): 277–310.

British Petroleum (BP). 2010a. Press Conference, BP Chief Executive Officer Tony Hayward, May 31.

———. 2010b. "Monitoring of Water Quality and Chemistry in Vicinity of the MC252 Oil Spill Location." Report of R/V Brooks McCall, May 7–26.

———. 2010c. "Monitoring of Water Quality and Chemistry in Vicinity of the MC252 Oil Spill Location." Report of R/V Ocean Veritas, June 1–5.

———. 2010d. Press Conference, BP Chief Operating Officer Doug Suttles, June 9.

Brown, Kate. 2013. *Plutopia: Nuclear Families, Atomic Cities, and the Great Soviet and American Plutonium Disasters.* Oxford: Oxford University Press.

———. 2016. "The Last Sink: The Human Body as the Ultimate Radioactive Storage Site." *Rachel Carson Center Perspectives,* no. 1, 41–48.

Brown, Phil, Rachel Morello-Frosch, and Stephen Zavestoski, eds. 2011. *Contested Illnesses: Citizens, Science, and Health Social Movements.* Berkeley: California University Press.

Brown, Wendy. 2019. *In the Ruins of Neoliberalism: The Rise of Antidemocratic Politics in the West.* New York: Columbia University Press.

Buck-Morss, Susan. 1977. *The Origins of Negative Dialectics.* New York: Free Press.

Bullard, Robert. 1993. "Race and Environmental Justice in the United States." *Yale Journal of International Law* 18(1): 319–35.

———. 1990. *Dumping in Dixie: Race, Class, and Environmental Quality.* New York: Routledge.

Burns, K. A., S. D. Garrity, and S. C. Levings. 1993. "How Many Years until Mangrove Ecosystems Recover from Catastrophic Oil Spills?" *Marine Pollution Bulletin* 26(5): 239–48.

Caldwell, Lynton K. 1963. "Environment: A New Focus for Public Policy?" *Public Administration Review* 23(3): 132–39.

———. 1964. "Biopolitics: Science, Ethics, and Public Policy." *Yale Review* 54(1): 1–16.

———. 1966. "Administrative Possibilities for Environmental Control." In *Future Environments of North America,* edited by F. Darling and J. Milton, 513–37. New York: Natural History Press.

———. 1970. *Environment: A Challenge to Modern Society.* Boston: Natural History Press.

———. 1998. *The National Environmental Policy Act: An Agenda for the Future.* Indianapolis: Indiana University Press.

Çaliskan, Koray, and Michel Callon. 2009. "Economization, Part 1: Shifting Attention from the Economy towards Processes of Economization." *Economy and Society* 38(3): 369–98.

———. 2010. "Economization, Part 2: A Research Programme for the Study of Markets." *Economy and Society* 39(1): 1–32.

Callon, Michel. 1989. "Society in the Making: The Study of Technology as a Tool for Sociological Analysis." In *The Social Construction of Technological*

Systems, edited by Wiebe Bijker, Thomas Hughes, and Trevor Pinch, 83–103. Cambridge, MA: MIT Press.

Callon, Michel, Pierre Lascoumes, and Yannick Barthe. 2009. *Acting in an Uncertain World: An Essay on Technical Democracy.* Cambridge, MA: MIT Press.

Camilli, Richard, Christopher Reddy, Dana Yoerger, Benjamin Van Mooy, Michael Jakuba, James Kinsey, Cameron McIntrye, Sean Sylva, and James Maloney. 2010. "Tracking Hydrocarbon Plume Transport and Biodegradation at Deepwater Horizon." *Science* 330(6001): 201–4.

Cangialosi, John, Andrew Latto, and Robbie Berg. 2018. "Tropical Cyclone Report: Hurricane Irma (30 August–12 September, 2017)." National Hurricane Center, NOAA.

Canguilhem, Georges. (1966) 1991. *The Normal and the Pathological.* Translated by Carolyn Fawcett. Cambridge, MA: Zone Books.

Cannon, Christopher, and Tiffany Kary. 2018. "Cancer-Linked Chemicals Manufactured by 3M Are Turning Up in Drinking Water." *Bloomberg Business News Online*, November 2. www.bloomberg.com/graphics/2018-3M-groundwater-pollution-problem/.

Cantrill, James G., and Christine L. Oravec. 1996. *Symbolic Earth: Discourse and Our Creation of the Environment.* Louisville: University Press of Kentucky.

Caribbean Development Bank. 2003. "The Current State and Future of Caribbean Agriculture." Caribbean Development Bank Report, January 3.

Carr, Donald. 1965. *The Breath of Life.* New York: Norton.

Carson, Rachel. 1962. *Silent Spring.* New York: Houghton Mifflin.

CBC News. 2014. "Oilsands Study Confirms Tailings Found in Groundwater, River." www.cbc.ca/news/canada/edmonton/oilsands-study-confirms-tailings-found-in-groundwater-river-1.2545089.

Cepek, Michael. 2012. "The Loss of Oil: Constituting Disaster in Amazonian Ecuador." *Journal of Latin American and Caribbean Anthropology* 17(3): 393–412.

———. 2016. "There Might Be Blood: Oil, Humility, and the Cosmopolitics of a Cofán Petro-Being." *American Ethnologist* 43(4): 623–35.

———. 2018. *Life in Oil: Cofán Survival in the Petroleum Fields of Amazonia.* Austin: Texas University Press.

Césaire, Aimé. 1990. *Lyric and Dramatic Poetry, 1946–82.* Translated by Clayton Eshleman and Annette Smith. Charlottesville: University of Virginia Press.

Chakrabarty, Dipesh. 2009. "The Climate of History: Four Theses." *Critical Inquiry* 35(2): 197–222.

Chakraborty, Jayajit. 2022. "Children's Exposure to Vehicular Pollution: Environmental Injustice in Texas, USA." *Environmental Research* 204(A): 34–41.

Chatterjee, Piya, Monisha Das Gupta, and Richard Rath. 2010. "Imperial Plantations: Past, Present, and Future Directions." *Journal of Historical Sociology* 23(1): 1–15.

Checker, Melissa. 2005. *Polluted Promises: Environmental Racism and the Search for Justice in a Southern Town.* New York: New York University Press.

————. 2007. "But I Know It's True: Environmental Risk Assessment, Justice, and Anthropology." *Human Organization* 66(2): 112–24.

Chen, Mel. 2007. "Racialized Toxins and Sovereign Fantasies." *Discourse* 29(2–3): 367–83.

————. 2011. "Toxic Animacies, Inanimate Affections." *GLQ: A Journal of Lesbian and Gay Studies* 17(2–3): 265–86.

Chen, Yiqun. 2009. "Cancer Incidence in Fort Chipewyan, Alberta 1995–2006." Alberta Cancer Board Report, February.

Chernick, Sidney. 1978. *The Commonwealth Caribbean: The Integration Experience.* Washington, DC: World Bank.

Choy, Timothy. 2005. "Articulated Knowledges: Environmental Forms after Universality's Demise." *American Anthropologist* 107(1): 5–18.

————. 2011. *Ecologies of Comparison: An Ethnography of Endangerment in Hong Kong.* Durham, NC: Duke University Press.

Cintrón, Gilberto, and Yara Schaeffer-Novelli. 1982. *Mangrove Forests: Ecology and Response to Natural and Man Induced Stressors.* Springfield, VA: US Department of Commerce, National Technical Information Service.

Clark, Lesley. 2010. "NOAA Ship to Study Underwater Oil Near Site of Leak," *Miami Herald*, June 2.

Clarke, Lee. 2005. *Worst Cases: Terror and Catastrophe in the Popular Imagination.* Chicago: University of Chicago Press.

Clifford, James. 1988. "Identity in Mashpee." In *The Predicament of Culture: Twentieth-Century Ethnography, Literature, and Art,* 277–348. Cambridge, MA: Harvard University Press.

Cohen, Lizabeth. 2004. *A Consumers Republic: The Politics of Mass Consumption in Postwar America.* New York: Vintage.

Cole, Luke, and Sheila Foster. 2001. *From the Ground Up: Environmental Racism and the Rise of the Environmental Justice Movement.* New York: New York University Press.

Coleman, A. L. 1955. "The Development of Threshold Limit Values." *AMA Archives of Industrial Health* 12(6): 685–87.

Coll, Steve. 2012. *Private Empire: Exxon-Mobil and American Power.* New York: Penguin Press.

Collier, Stephen. 2008. "Enacting Catastrophe: Preparedness, Insurance, Budgetary Rationalization." *Economy and Society* 37(2): 224–50.

Collier, Stephen, and Andrew Lakoff. 2021. *The Government of Emergency: Vital Systems, Expertise, and the Politics of Security.* Princeton, NJ: Princeton University Press.

Commoner, Barry. 1958. "The Fallout Problem." *Science* 127(3305): 1023–26.

————. 1967. *Science and Survival.* New York: Viking.

————. 1970. "Kick-off Rally: Teach-in on the Environment." Keynote address, presented at University of Michigan, Ann Arbor, March 11).

————. 1971. *The Closing Circle: Nature, Man, and Technology.* New York: Knopf.

Cons, Jason. 2019. "Delta Temporalities: Choked and Tangled Futures in the Sunbarbans." *Ethnos*, 1–22.

Cook, W. 1945. "Maximum Allowable Concentrations of Industrial Contaminants." *Industrial Medicine* 14(11): 936–46.

Coole, Diana, and Samantha Frost. 2010. *New Materialisms: Ontology, Agency, and Politics*. Durham, NC: Duke University Press.

Coronil, Fernando. 1997. *The Magical State: Nature, Money, and Modernity in Venezuela*. Chicago: University of Chicago Press.

Corredor, Jorge, Julio Morell, and Carlos Castillo. 1990. "Persistence of Spilled Crude Oil in a Tropical Intertidal Environment." *Marine Pollution Bulletin* 21(8): 385–88.

Costanza, Robert, Ralph d'Arge, Rudolf de Groot, Stephen Farber, Monica Grasso, Bruce Hannon, Karin Limburg, Shahid Naeem, Robert V. O'Neill, Jose Paruelo, Robert G. Raskin, Paul Sutton, and Marjan van den Belt. 1997. "The Value of the World's Ecosystem Services and Natural Capital." *Nature* 387 (May 15): 253–60.

Coulthard, Glen. 2014. *Red Skin, White Masks: Rejecting the Colonial Politics of Recognition*. Minneapolis: University of Minnesota Press.

Council of Environmental Quality (CEQ). 1979. *Environmental Quality: The Tenth Annual Report of the Council on Environmental Quality*. Executive Office of the President, Council on Environmental Quality.

Couwels, John. 2010. "New Oil Plume Evidence Uncovered." CNN, June 7.

CPSA (College of Physicians and Surgeons of Alberta). 2009. "Investigation Report: Dr. John O'Connor." Report. Edmonton, AB.

Crawford, Neta. 2019. "Pentagon Fuel Use, Climate Change, and the Costs of War." Report, Costs of War Project, Watson Institute, Brown University. November 13. https://watson.brown.edu/costsofwar/files/cow/imce/papers /Pentagon%20Fuel%20Use%2C%20Climate%20Change%20and%20 the%20Costs%20of%20War%20Revised%20November%202019%20 Crawford.pdf.

Crosby, Alfred. 2006. *Children of the Sun: A History of Humanity's Unappeasable Appetite for Energy*. New York: Norton.

Crutzen, Paul. 1987. "Climatic Effects of Nuclear War." In *Effects of Nuclear War on Health and Health Services*, 65–82. Report of the World Health Organization.

———. 2002. "Geology of Mankind." *Nature* 415 (January 3): 23.

Crutzen, Paul, and J. W. Birks. 1982. "The Atmosphere After a Nuclear War: Twilight at Noon." *Ambio* 11: 114–25.

Darling, Fraser, and John Milton, eds. 1966. *Future Environments of North America: Transformation of a Continent*. New York: Natural History Press.

Das, Veena. 1995. *Critical Events: An Anthropological Perspective on Contemporary India*. Oxford: Oxford University Press.

Daston, Lorraine, and Peter Galison. 2007. *Objectivity*. Cambridge, MA: Zone Books.

Dave, Naisargi. 2012. *Queer Activism in India: A Story in the Anthropology of Ethics*. Durham, NC: Duke University Press.

Davis, J. H., Jr. 1938. "Mangroves: Makers of Land." *Nature* 31:551–53.

Davis, Mike. 1986. *Prisoners of the American Dream: Politics and Economy in the History of the American Working Class*. New York: Verso.

———. 2002. *Dead Cities*. New York: New Press.

———. 2006. *Planet of Slums*. New York: Verso.

de Gouw, J. A., A. Middlebrook, C. Warneke, R. Ahmadov, E. Atlas, R. Bahreini, D. Blake, C. Brock, J. Brioude, D. Fahey, F. Fehsenfeld, J. Holloway, M. Henaff, R. Lueb, S. McKeen, J. Meagher, D. Murphy, C. Paris, D. Parrish, A. Perring, I. Pollack, A. Ravishankara, A. Robinson, T. Ryerson, J. Schwarz, J. Spackman, A. Srinivasan, and L. Watts. 2011. "Organic Aerosol Formation Downwind from the Deepwater Horizon Oil Spill." *Science* 331(6022): 1295–99.

De Jongh, John, Jr. 2012. "The U.S. Virgin Islands: Economic Impact of the HOVENSA Closing." Presentation by governor to Federal Interagency Meeting, February 24.

de la Cadena, Marisol. 2015. *Earth Beings: Ecologies of Practice Across Andean Worlds*. Durham, NC: Duke University Press.

Dean, Warren. 1997. *With Broadax and Firebrand: The Destruction of the Brazilian Atlantic Forests*. Berkeley: University of California Press.

Deger, Leylâ, Céline Plante, Louis Jacques, Sophie Goudreau, Stéphane Perron, John Hicks, Tom Kosatsky, and Audrey Smargiassi. 2012. "Active and Uncontrolled Asthma among Children Exposed to Air Stack Emissions of Sulphur Dioxide from Petroleum Refineries in Montreal, Quebec: A Cross-Sectional Study." *Canadian Respiratory Journal* 19(2): 97–102.

DePass, Dee. 2019. "Rash of Lawsuits Intensify Concerns of 3M Liability over PFAS Chemicals." *Star Tribune*, December 23.

der Sluijs, Jeron van, Josee van Eijndhoven, Simon Shackley, and Brian Wynne. 1998. "Anchoring Devices in Science for Policy: The Case of Consensus around Climate Sensitivity." *Social Studies of Science* 28(2): 291–23.

DeSouza, Mike. 2018. "Cleaning up Alberta's Oilpatch Could Cost $260 Billion, Internal Documents Warn." *National Observer*, November 21.

Di Chiro, Giovanna. 2010. "Polluted Politics? Confronting Toxic Discourse, Sex Panic, and Eco-Normativity." In *Queer Ecologies*, edited by Catriona Mortimer-Sandilands, and Bruce Erickson, 199–234. Indianapolis: Indiana University Press.

Dietz, James. 1986. *Economic History of Puerto Rico: Institutional Change and Capitalist Development*. Princeton, NJ: Princeton University Press.

Dodge, Richard E., Thomas Ballou, Stephen Hess, Anthony Knap, and Thomas Sleeter. 1995. *The Effects of Oil and Chemically Dispersed Oil in Tropical Ecosystems: 10 Years of Monitoring Experimental Sites*. MSRC Technical Report Series 95-104. Washington, DC: Marine Spill Response Corporation.

Dolan, Catherine, and Dinah Rajak, eds. 2016. *The Anthropology of Corporate Social Responsibility*. New York: Berghahn.

DuBois, W. E. B. 1915. "The African Roots of War." *Atlantic*, May, 4–7.

Dubos, Rene J., and Barbra Ward. 1972. *Only One Earth: The Care and Maintenance of a Small Planet*. New York: W. W. Norton.

Duke, Norman, Jan-Olaf Meynecke, Sabine Dittmann, A.M. Ellison, Klaus Anger, Uta Berger, Stefano Cannicci, K. Diele, Katherine Ewel, Colin Field, Nico Koedam, Shing Yip Lee, Cyril Marchand, Inga Nordhaus, and

Farid Dahdouh-Guebas. 2007. "A World without Mangroves?" *Science* 317(5834): 41–42.

Duke, Norman C., Zuleika S. Pinzon, and Martha C. Prada. 1997. "Large-Scale Damage to Mangrove Forests Following Two Large Oil Spills in Panama." *Biotropica* 29(1): 2–14.

Dunaway. 2015. *Seeing Green: The Uses and Abuses of American Environmental Images.* Chicago: Chicago University Press.

Dunlap, Thomas. 2008. *DDT, Silent Spring, and the Rise of Environmentalism.* Seattle: University of Washington Press.

Eckelmann, Walter, Laurence Kulp, and Arthur Schulert. 1958. "Strontium-90 in Man, II." *Science* 127(3293): 266–74.

Eden, Lynn. 2004. *Whole World on Fire: Organizations, Knowledge, and Nuclear Weapons Devastation.* Ithaca, NY: Cornell University Press.

"Editorial: IS Imperial Ambitions and Iraq." 2002. *Monthly Review* 54(7): 1–13.

Edsall, David. 1918. "Medical-Industrial Relations of the War," *John Hopkins Hospital Bulletin* (September): 197–205.

Edwards, Jocelyn. 2014. "Oil Sands Pollutants in Traditional Foods." *Canadian Medical Association Journal* 186(12): E444.

Edwards, Paul N. 2010. *A Vast Machine: Computer Models, Climate Data, and the Politics of Global Warming.* Cambridge, MA: MIT Press.

Egan, Michael. 2007. *Barry Commoner and the Science of Survival: The Remaking of American Environmentalism.* Cambridge, MA: MIT Press.

Eggertson, Laura. 2009. "High Cancer Rates among Fort Chipewyan Residents." *Canadian Medical Association Journal* 181(12): E309.

Ekirch, Arthur. 1963. *Man and Nature in America.* New York: Columbia University Press.

Elkins, H. 1948. "Maximum Allowable Concentrations of Mixtures." *American Industrial Hygiene Association Journal* 23: 132–36.

Ellison, Aaron, and Elizabeth Farnsworth. 1996. "Anthropogenic Disturbance of Caribbean Mangrove Ecosystems: Past Impacts, Present Trends, and Future Predictions." *BioTropica* 28(4a): 549–65.

Energy Information Administration (EIA). 2009. *Caribbean.* Country Analysis Briefs. Washington, DC: Department of Energy.

———. 2016a. *U.S. Virgin Islands.* Country Analysis Briefs. Washington, DC: Department of Energy.

———. 2016b. *Puerto Rico.* Country Analysis Briefs. Washington, DC: Department of Energy.

———. 2021. "Oil and Petroleum Products Explained: Imports and Exports." Report, April 13. Washington, DC: Department of Energy.

Engel, Antke, and Renate Lorenz. 2013. "Toxic Assemblages, Queer Socialities: A Dialogue of Mutual Poisoning." *e-flux* 44 (April). www.e-flux.com/journal/44/60173/toxic-assemblages-queer-socialities-a-dialogue-of-mutual-poisoning/.

Environmental Protection Agency (EPA). 2011. "News Release: HOVENSA LLC, Clean Air Act Settlement." Washington, DC, January 26.

———. 2017. "RE: Limetree Bay Terminals." Letter from EPA Caribbean Environmental Protection Division Director Carmen Perez to Chief, Antilles Regulatory Section, US Corp of Engineers Sindulfo Castillo, December 29.

———. 2019. "News Release: EPA Awards $412,101 to the U.S. Virgin Islands to Improve Air Quality." Washington, DC, November 12.

———. 2021 *PFAS Testing Strategy: Identification of Candidate Per- and Polyfluoroalkyl Substances (PFAS) for Testing*. Report, October 2021. www.epa.gov/system/files/documents/2021-10/pfas-natl-test-strategy.pdf.

Erikson, Kai. 1976. *Everything in Its Path: Destruction of Community in the Buffalo Creek Flood*. New York: Simon and Schuster.

Escobar, Arturo. 2008. *Territories of Difference: Place, Movements, Life, Redes*. Durham, NC: Duke University Press.

Esposito, John. 1970. *Vanishing Air: Ralph Nadar's Study Group Report on Air Pollution*. New York: Grossman.

Estes, Nick. 2019. *Our History Is the Future*. Brooklyn: Verso.

Estes, Nick, and Roxanne Dunbar-Ortiz. 2020. "Examining the Wreckage," *Monthly Review* 72(3): 4–12.

Fabian, Johannes. 1983. *Time and the Other: How Anthropology Makes Its Object*. New York: Columbia University Press.

Farmer, Paul. 2001. *Infections and Inequalities: The Modern Plagues*. Berkeley: University of California Press.

Fassin, Didier. 2009. "Another Politics of Life Is Possible." *Theory, Culture & Society* 26(5): 44–60.

———. 2012. *Humanitarian Reason: A Moral History of the Present*. Berkeley: University of California Press.

Feller, I. C., C. E. Lovelock, U. Berger, K. L. McKee, S. B. Joyce, and M. C. Ball. 2010. "Biocomplexity in Mangrove Ecosystems." *Annual Review of Marine Science* 2:395–417.

Ferguson, James. 2005. "Seeing Like an Oi Company: Space, Security, and Global Capital in Neoliberal Africa." *American Anthropologist* 107(3): 377–82.

Fisherman, Robert. 1987. *Bourgeois Utopias: The Rise and Fall of Suburbia*. New York: Basic Books.

Fitzgerald, Deborah. 2003. *Every Farm a Factory: The Industrial Ideal in American Agriculture*. New Haven, CT: Yale University Press.

Forsyth, Tim, and Andrew Walker. 2008. *Forest Guardians, Forest Destroyers: The Politics of Environmental Knowledge in Northern Thailand*. Seattle: University of Washington Press.

Fort McKay Industrial Relations Corporation (FMIRC). 2010. "Cultural Heritage Assessment Baseline: Pre-development (1960s) to Current (2008)." Report, March.

Fortun, Kim. 2001. *Advocacy after Bhopal: Environmentalism, Disaster, New Global Orders*. Chicago: Chicago University Press.

———. 2012. "Ethnography in Late Industrialism." *Cultural Anthropology* 27(3): 446–64.

———. 2014. "From Latour to Late Industrialism." *Hau* 4(1): 309–29.

Foster, John Bellamy. 2000. *Marx's Ecology: Materialism and Nature*. New York: Monthly Review Press.

———. 2009. "The Absolute General Law of Environmental Degradation under Capitalism." *Capitalism, Nature, Socialism* 3(3): 77–81.

Fowler, John, ed. 1960. *Fallout: A Study of Superbombs, Strontium-90, and Survival.* New York: Basic Books.

Frank, Richard, James W. Roy, Greg Bickerton, Steve J. Rowland, John V. Headley, Alan G. Scarlett, Charles E. West, Kerry M. Peru, Joanne L. Parrott, F. Malcolm Conly, and L. Mark Hewitt. 2014. "Profiling Oil Sands Mixtures from Industrial Developments and Natural Groundwaters for Source Identification." *Environmental Science & Technology* 48(5): 2660–70.

Freeman, Carla. 2000. *High Tech and High Heels in the Global Economy: Women, Work, and Pink-Collar Identities in the Caribbean.* Durham, NC: Duke University Press.

Fumoleau, Rene. 2004. *As Long as This Land Shall Last: History of Treaty 8 and Treaty 11, 1870–1939.* Calgary, AB: University of Calgary Press.

Garrity, Stephen, Sally Levings, and Kathryn Burns. 1994. "The Galeta Oil Spill: Longterm Effects on the Physical Structure of the Mangrove Fringe." *Estuarine, Coastal, and Shelf Science* 38(4): 327–48.

Getter, Charles, Geoffrey I. Scott, and Jacqueline Michel. 1981. "Effects of Oil Spills on Mangrove Forests: A Comparison of Five Oil Spill Sites in the Gulf of Mexico and the Caribbean Sea." *International Oil Spill Proceedings* no. 1, 535–40.

Gilfilian, Edward, David Page, Ray Gerber, Sherry Hansen, Judy Cooley, and Janet Hotham. 1981. "Fate of the Zoe Colocotroni Oil Spill and Its Effects on Infaunal Communities Associated with Mangroves." *International Oil Spill Proceedings* no. 1, 353–60.

Gillis, Justin. 2010a. "Giant Plumes of Oil Forming Under the Gulf." *New York Times,* May 16.

———. 2010b. "Scientists Build Case for Undersea Plumes." *New York Times,* May 28.

———. 2010c. "A Proliferation of Plumes?" *Green Blog, New York Times,* June 2. http:green.blog.nytimes.com/2010/06/02.

Glissant, Édouard. (1981) 1999. *Caribbean Discourse: Selected Essays.* Translated by J. M. Dash. Charlottesville: University of Virginia Press.

Goldman, Mara, Paul Nadasdy, and Matthew Turner. 2011. *Knowing Nature: Conversations at the Intersection of Political Ecology and Science Studies.* Chicago: University of Chicago Press.

Golley, Frank. 1993. *A History of the Ecosystem Concept in Ecology: More Than the Sum of the Parts.* New Haven, CT: Yale University Press.

Gordillo, Gaston. 2014. *Rubble.* Durham, NC: Duke University Press.

Gorman, Hugh. 2001. *Refining Efficiency: Pollution Concerns, Regulatory Mechanisms, and Technological Change in the U.S. Petroleum Industry.* Athens: University of Ohio Press.

Gottlieb, Robert. 1993. *Forcing the Spring: The Transformation of the American Environmental Movement.* Washington, DC: Island Press.

Grandjean, Philippe. 2017. "Expert Report of Philippe Grandjean, MD." Prepared on Behalf of Plaintiff State of Minnesota, Civil Action No. 27-CV-10-28862, State of Minnesota et al. v. 3M Company, September 22.

Grant, D. A., P. G. Clarke, and W. G. Allaway. 1993. "The Response of Grey Mangrove Seedlings to Spills of Crude Oil." *Journal of Experimental Marine Biology and Ecology* 171(2): 273–95.

Grove, Richard. 1995. *Green Imperialism: Colonial Expansion, Tropical Island Edens, and the Origins of Environmentalism, 1600–1860.* Cambridge: Cambridge University Press.

Grover, Herbert, and Mark Harwell. 1985. "Biological Effects of Nuclear War II: Impact on the Biosphere." *BioScience* 35(9): 576–83.

Guha, Ramachandra. 2000. *Environmentalism: A Global History.* New York: Longman.

Gundlach, Erich, and Miles O. Hayes. 1978. "Classification of Coastal Environments in Terms of Potential Vulnerability to Oil Spill Impact." *Marine Technology Society Journal* 12(1): 18–27.

Gundlach, Erich, Miles O. Hayes, and Charles Getter. 1979. "Determining Environmental Protection Priorities in Coastal Ecosystems." In *Proceedings of 1979 U.S. Fish and Wildlife Service Pollution Response Workshop.* St. Petersburg, FL: US Fish and Wildlife Service.

Gunewardena, Nandini, and Mark Scholler, eds. 2008. *Capitalizing on Catastrophe: Neoliberal Strategies in Disaster Reconstruction.* New York: AltaMira.

Haddad, Robert, and Steven Murawski. 2010. "Analysis of Hydrocarbons in Samples Provided from the Cruise of the R/V Weatherbird II, May 23–26, 2010." NOAA Report.

Hamblin, Jacob. 2013. *Arming Mother Nature: The Birth of Catastrophic Environmentalism.* Oxford: Oxford University Press.

Hamilton, Alice. 1918. "Trinitrotoluene Poisoning." *Monthly Labor Review,* no. 7, 236–51.

———. 1919. "Practical Points in the Prevention of TNT Poisoning." *Monthly Labor Review,* no. 8, 248–72.

———. 1943. *Exploring the Dangerous Trades: The Autobiography of Alice Hamilton, M.D.* Boston: Little Brown.

Hanieh, Adam. 2021. "Petrochemical Empire: The Geo-Politics of Fossil-Fuelled Production," *New Left Review* 130 (July–August): 25–51.

Hansen, James. 2011. "Silence Is Deadly." Public letter. www.columbia.edu /~jeh1/mailings/2011/20110603_SilenceIsDeadly.pdf.

Haraway, Donna J. 2016. *Staying with the Trouble: Making Kin in the Chthulucene.* Durham. NC: Duke University Press.

Harris, Marvin. 1979. *Cultural Materialism: The Struggle for a Science of Culture.* New York: Vintage.

Hartog, Johannes. 1968. *Curaçao, from Colonial Dependence to Autonomy.* Oranjestad, Aruba: De Wit.

Harvey, David. 1996. *Justice, Nature, and the Geography of Difference.* Oxford: Blackwell.

———. 2003. *The New Imperialism.* Oxford: Oxford University Press.

———. 2007. *A Brief History of Neoliberalism.* Oxford: Oxford University Press.

——. 2014. *Seventeen Contradictions and the End of Capitalism*. Oxford: Oxford University Press.

Hayes, T. 1977. "Sinking of Tanker St. Peter off Columbia." *International Oil Spill Conference Proceedings* 1:289–91.

Hays, Samuel. 1959. *Conversation and the Gospel of Efficiency: The Progressive Conservation Movement, 1890–1920*. Pittsburgh: University of Pittsburgh Press.

——. 1989. *Beauty, Health, and Permanence: Environmental Politics in the United States, 1955–1985*. Cambridge: Cambridge University Press.

——. 2000. *A History of Environmental Politics since 1945*. Pittsburgh: University of Pittsburgh Press.

Hearing on Environmental Quality. 1969. Congressional Committee on Natural Resources, US House of Representatives (testimony of Michael McCloskey for the Sierra Club, April 16).

Hearing on Greenhouse Effect and Global Climate Change. 1988. Congressional Committee on Energy and Natural Resources, US Senate (statement of Dr. James Hansen, Director, NASA Goddard Institute for Space Studies, June 23).

Hecht, Gabrielle. 2018. "Residue." Somatosphere, January 8. http://somatosphere.net/2018/residue.html/.

Helmreich, Stefan. 2009. *Alien Oceans: Anthropological Voyages in Microbial Seas*. Berkeley: University of California Press.

Herty, Charles. 1916. "The Expanding Relations of Chemistry in America," *Science*, no. 44, 475–82.

Hetherington, Kregg. 2019. *Infrastructure, Environment and Life in the Anthropocene*. Durham, NC: Duke University Press.

Hewitt, Kenneth. 1983. "The Idea of Calamity in a Technocratic Age." In *Interpretations of Calamity*, edited by Kenneth Hewitt, 2–32. London: Allen & Unwin.

Hiar, Corbin. 2019. "Trump Administration Provides 'Customer' Service to Troubled Refinery." *E&E News*, November 21.

Hickel, Walter. 1971. *Who Owns America?* Englewood Cliffs, NJ: Prentice-Hall.

Hoch, Myron. 1948. "The Oil Strike of 1945." *Southern Economic Journal* 15(2): 117–33.

Hoff, Rebecca. 2014. *Oil Spills in Mangroves: Planning and Response Considerations*. Washington, DC: National Oceanic and Atmospheric Administration.

Holder, Eric. 2012. "Attorney General Eric Holder Speaks at the BP Press Conference," November 15. www.justice.gov/opa/speech/attorney-general-eric-holder-speaks-bp-press-conference.

Holmes, Morgan. 2000. "Queer Cut Bodies." In *Queer Frontiers: Millennial Geographies, Genders and Generations*, edited by J. Boone, 84–110. Madison: Wisconsin University Press.

Holmes, Seth. 2013. *Fresh Fruit, Broken Bodies: Migrant Farmworkers in the United States*. Berkeley: California University Press.

Home Journal. 1964. "Governor Plans to Wipe Out St. Croix Feudal System." June 30, 1.

Horkheimer, Max, and Theodor Adorno. (1947) 2002. *Dialectic of Enlightenment: Philosophical Fragments.* Edited by Gunzelin Noerr. Translated by Edmund Jephcott. Stanford, CA: Stanford University Press.

Horowitz, Tony. 2014. *Boom: Oil, Money, Cowboys, Strippers, and the Energy Rush That Could Change America Forever.* Seattle: Amazon Kindle Singles.

Hornberg, Alf. 1998. "Towards an Ecological Theory of Unequal Exchange: Articulating World System Theory and Ecological Economics." *Ecological Economics* 25(1): 127–36.

Horton, Sarah. 2016. *They Leave Their Kidneys in the Fields: Illness, Injury, and Illegality among U.S. Farmworkers.* Berkeley: California University Press.

Hotelling, Harold. 1931. "The Economics of Exhaustible Resources." *Journal of Political Economy* 39(2): 137–75.

Hrudey, Steve. 2011. "Oil Sands Contaminants." *Environmental Health Perspectives* 119(8): A330.

Huber, Matthew. 2013. *Lifeblood: Oil, Freedom, and the Forces of Capital.* Minneapolis: University of Minnesota Press.

Hueper, W. C., and Paul Kotin. 1955. "Relationship of Industrial Carcinogens to Cancer in the General Population." *Public Health Report* 70 (3): 331–34

Hughes, David. 2013. "Climate Change and the Victim Slot: From Oil to Innocence." *American Anthropologist* 115(4): 570–81.

Hurley, Andrew. 1995. *Environmental Inequalities: Class, Race, and Industrial Pollution in Gary, Indiana, 1945–1980.* Chapel Hill: University of North Carolina Press.

Immerwahr, Daniel. 2019. *How to Hide an Empire: A History of the Greater United States.* New York: Picador.

"Industrial Poisons and Diseases." 1918. *Monthly Labor Review*, July, 185-92.

Inglehart, Ronald. 1981. "Post-Materialism in an Environment of Insecurity." *American Political Science Review* 75(4): 880–900.

Intergovernmental Panel on Climate Change (IPCC). 2014. *Climate Change 2014: Synthesis Report; Contribution of Working Groups I, II and III to the Fifth Assessment Report of the Intergovernmental Panel on Climate Change.* [Edited by R. K. Pachauri and L. A. Meyer.] Geneva: IPCC.

———. 2018. *2018: Global Warming of 1.5°C: An IPCC Special Report on the Impacts of Global Warming of 1.5°C above Pre-industrial Levels and Related Global Greenhouse Gas Emission Pathways, in the Context of Strengthening the Global Response to the Threat of Climate Change, Sustainable Development, and Efforts to Eradicate Poverty.* [Edited by V. Masson-Delmotte, P. Zhai, H. O. Pörtner, D. Roberts, J. Skea, P.R. Shukla, A. Pirani, W. Moufouma-Okia, C. Péan, R. Pidcock, S. Connors, J. B. R. Matthews, Y. Chen, X. Zhou, M. I. Gomis, E. Lonnoy, T. Maycock, M. Tignor, and T. Waterfield]. Geneva: IPCC.

Jackson, Kenneth. 1985. *Crabgrass Frontier: The Suburbanization of the United States.* Oxford: Oxford University Press.

Jacobs, Meg. 2016. *Panic at the Pump: The Energy Crisis and the Transformation of American Politics in the 1970s.* New York: Hill and Wang.

Jain, S. Lochlann. 2007. "Cancer Butch." *Cultural Anthropology* 22, no. 4 (November): 501–38.

James, C. L. R. (1939) 1989. *The Black Jacobins: Toussaint L'Ouverture and the San Domingo Revolution.* Rev. ed. New York: Vintage.

Jasanoff, Sheila. 1984. *Learning from Disaster: Risk Management after Bhopal.* Philadelphia: University of Pennsylvania Press.

———. 1990. *The Fifth Branch: Science Advisors as Policymakers.* Cambridge, MA: Harvard University Press.

Johannes, R. E., and Susan B. Betzer. 1975. "Introduction: Marine Communities Respond Differently to Pollution in the Tropics than at Higher Latitudes." In *Tropical Marine Pollution*, edited by E. J. Furguson Wood and R. E. Johannes, 1–12. Elsevier Oceanography Series. Amsterdam: Elsevier Publishing.

Johnson, E. A., and K. Miyanishi. 2008. "Creating New Landscapes and Ecosystems: The Alberta Oil Sands." *New York Academy of Sciences* 1134: 120–45.

Johnson, Tracy. 2016. "Alberta Energy Regulator Tries to Stem Tide of Orphan Wells." CBC News/Radio Canada, June 21. www.cbc.ca/news/canada/calgary/alberta-orphan-wells-aer-rule-change-1.3646235.

Johnston, Barbara, and Holly Barker. 2008. *Consequential Damages of Nuclear War: The Rongelap Report.* Walnut Creek, CA: Left Coast Press.

Joint Oil Sands Monitoring Program (JOSM). 2016. "Joint Oil Sands Monitoring Program Emissions Inventory Report." Report submitted to Government of Canada, April.

Joly, Tara, Hereward Longley, Carmen Wells, and Jenny Gerbrandt. 2018. "Ethnographic Refusal in Traditional Land Use Mapping: Consultation, Impact Assessment, and Sovereignty in the Athabasca Oil Sands Region." *Extractive Industries and Society* 5:335–43.

Jones, Charles O. 1975. *Clean Air: The Policies and Politics of Pollution Control.* Pittsburgh: Pittsburgh University Press.

Jones, Toby. 2014. "Toxic War and the Politics of Uncertainty in Iraq." *International Journal of Middle East Studies* 46(4): 797–99.

Joye, Samantha, Ira Leifer, Ian MacDonald, Jeffrey Chanton, Christof Meile, Andreas Teske, Joel Kostka, Ludmila Chistoserdova, Richard Coffin, David Hollander, Miriam Kastner, Joseph Montoya, Gregor Rehder, Evan Solomon, Tina Treude, and Tracy Villareal. 2011. Comment on "A Persistent Oxygen Anomaly Reveals the Fate of Spilled Methane in the Deep Gulf of Mexico." *Science* 32(6033): 1033.

Joye, Samantha, Ian MacDonald, Ira Leifer, and Vernon Asper. 2011. "Magnitude and Oxidation Potential of Hydrocarbon Gases Released from the BP Oil Well Blowout." *Nature Geoscience* 4:160–64.

Kelly, Erin, David W. Schindler, Peter V. Hodson, Jeffrey W. Short, Roseanna Radmanovich, and Charlene C. Nielsen. 2010. "Oil Sands Development Contributes Elements Toxic at Low Concentrations to the Athabasca River and Its Tributaries." *Proceedings of the National Academy of Sciences* 107(37): 16178–83.

Kelly, Erin, Jeffrey W. Short, David W. Schindler, Peter V. Hodson, Mingsheng Ma, Alvin K. Kwan, and Barbra L. Fortin. 2009. "Oil Sands Development Contributes Polycyclic Aromatic Compounds to the Athabasca River and Its Tributaries." *Proceedings of the National Academy of Sciences* 106(52): 22346–51.

Kennan, George F. 1970. "To Prevent a World Wasteland: A Proposal." *Foreign Affairs* 48(3): 191–203.

Kessler, John, David Valentine, Molly Redmond, Mengran Du, Eric Chan, Stephanie Mendes, Erik Quiroz, Christie Villanueva, Stephani Shusta, Lindsay Werra, Shari YvonLewis, and Thomas Weber. 2010. "A Persistent Oxygen Anomaly Reveals the Fate of Spilled Methane in the Deep Gulf of Mexico." *Science* 331(6015): 312–15.

Khayyat, Munira. 2022. *A Landscape of War: Resistant Ecologies in South Lebanon.* Berkeley: University of California Press.

Kirsch, Stuart. 2014. *Mining Capitalism: The Relationship between Corporations and Their Critics.* Berkeley: University of California Press.

Kirksey, Eben. 2015. *Emergent Ecologies.* Durham, NC: Duke University Press.

———. 2017. "Caring as Chemo-Ethnographic Method." Member Voices, *Fieldsights*, November 20. https://culanth.org/fieldsights/caring-as-chemo -ethnographic-method.

Kirksey, Eben, and Nicholas Shapiro. 2017. "Chemo-ethnography: An Introduction." *American Ethnologist* 32(4): 481–93.

Kjerfve, Björn, Luiz Drude de la Lacerda, and El Hadji Salif Diop, eds. 1997. *Mangrove Ecosystem Studies in Latin America and Africa.* UNESCO Report. Paris: UNESCO.

Klein, Naomi. 2007. *The Shock Doctrine: The Rise of Disaster Capitalism.* New York: Picador.

Kloppenburg, Jack Ralph. 1988. *First the Seed: The Political Economy of Plant Biotechnology.* Madison: University of Wisconsin Press.

Kohn, Eduardo. 2013. *How Forests Think: Toward an Anthropology beyond the Human.* Berkeley: University of California Press.

Kolbert, Elizabeth. 2006. "The Darkening Sea." *New Yorker*, November 20, 37–43.

Kotin, Paul, and W. C. Hueper. 1955. "Relationship of Industrial Carcinogens to Cancer in the General Population." *Public Health Reports* 70(3): 331–34.

Kroll-Smith, Steve, and Worth Lancaster. 2002. "Bodies, Environments, and a New Style of Reasoning." *Annals of the American Academy of Political and Social Science* 584(1): 203-12.

Krupnik, Igor, and Dyanna Jolly, eds. 2002. *The Earth Is Faster Now: Indigenous Observations of Arctic Environmental Change.* Fairbanks, AK: Arctic Research Consortium of the United States.

Kulp, Laurence and Arthur Schulert. 1962. "Strontium-90 in Man V." *Science* 136(3516): 619–32.

Kulp, Laurence, Arthur Schulert, and Walter Eckelmann. 1957. "Strontium-90 in Man." *Science* 125(3241): 219–25.

Kulp, Laurence, Arthur Schulert, and Elizabeth Hodges. 1959. "Strontium in Man III." *Science* 129(3358): 1249–55.

———. 1960. "Strontium in Man IV." *Science* 132(3425): 448–54.

Kunstler, James. 2005. *The Long Emergency.* New York: Atlantic Monthly Press.

Lacerda, L. D., J. E. Conde, C. Alarcon, R. Alvarex-León, P. R. Bacon, L. D'Croz, B. Kjerfve, J. Polaina, and M. Vannucci. 1993. "Mangrove Ecosystems of Latin America and the Caribbean: A Summary." In *Conservation and Sustainable Utilization of Mangrove Forests in Latin America and Africa Regions*, part I, *Latin America*, edited by L. D. Lacerda and S. Diop, 245–72. Wageningen, Netherlands: International Society for Microbial Ecology /ITTO.

Lahsen, Myanna. 2009. "A Science-Policy Interface in the Global South: The Politics of Carbon Sinks and Science in Brazil." *Climate Change* 97:339–72.

Lakoff, Andrew, ed. 2010. *Disaster and the Politics of Intervention.* New York: Columbia University Press.

Lamoreaux, Janelle. 2020. "Toxicology and the Chemistry of Cohort Kinship." Somatosphere, January. http://somatosphere.net/2020/chemical-kinship.html/.

Latour, Bruno. 1993. *We Have Never Been Modern.* Translated by Catherine Porter. Cambridge, MA: Harvard University Press.

———. 2004. *The Politics of Nature: How to Bring the Sciences into Democracy.* Cambridge, MA: Harvard University Press.

———. 2018. *Down to Earth: Politics in the New Climatic Regime.* Cambridge, UK: Polity Press.

Latour, Bruno, and Steve Woolgar. (1979) 1986. *Laboratory Life: The Construction of Scientific Facts.* Beverly Hills, CA: Sage Publications.

Lemke, Thomas. 2011. *Bio-Politics: An Advanced Introduction.* New York: New York University Press.

Lenin, Vladamir. 1920. *Imperialism: The Highest Stage of Capitalism.* London: International Publishers.

Lepore, Jill. 2017. "The Atomic Origins of Climate Science." *New Yorker*, January 22, 17–25.

Lerner, Sharon. 2015. "The Teflon Toxin." The Intercept August. https://theintercept.com/series/the-teflon-toxin/.

———. 2018. "3M Knew about the Dangers of PFOA and PFOS Decades Ago, Internal Documents Show." *Intercept*, July 31. https://theintercept.com/2018/07/31/3m-pfas-minnesota-pfoa-pfos/.

Lerner, Steve. 2010. *Sacrifice Zones.* Cambridge, MA: MIT Press.

Lewis, Roy. 1983. "Impact of Oil Spills on Mangrove Forests." In *Biology and Ecology of Mangroves.* edited by H. J. Teas, 171-84. The Hague: Springer Netherlands.

Lewis, Sir W. Arthur. 1950. *Industrial Development in the Caribbean.* Port of Spain: Caribbean Commission.

———. 1954. "Economic Development with Unlimited Supplies of Labor." *Manchester School* 22(2): 139–91.

Li, Fabiana. 2015. *Unearthing Conflict: Corporate Mining, Activism, and Expertise in Peru.* Durham, NC: Duke University Press.

Liboiron, Max. 2017. "Pollution Is Colonialism." Discard Studies, September. https://discardstudies.com/2017/09/01/pollution-is-colonialism/.

Limbert, Mandana. 2010. *In the Time of Oil: Piety, Memory, and Social Life in an Omani Town.* Stanford, CA: Stanford University Press.

Linnitt, Carol. 2014. "Alarming New Study Finds Contamination in Animals Downstream of Oilsands." *The Narwhal*, July 7.

Liroff, Richard. 1976. *A National Policy for the Environment: NEPA and Its Aftermath*. Bloomington: Indiana University Press.

Little, Paul. 1999. "Environments and Environmentalisms in Anthropological Research: Facing a New Millennium." *Annual Review of Anthropology* 28:253–84.

Little, Peter. 2022. *Burning Matters: Life, Labor, and E-Waste Pyropolitics in Ghana*. Oxford: Oxford University Press.

Llewellyn, Lynn G., and Clare Peiser. 1973. *National Environmental Policy Act (NEPA) and the Environmental Movement: A Brief History*. Report, Environmental Studies Division, Office of Research and Monitoring. Washington DC: Environmental Protection Agency.

Lock, Margaret. 2019. "Toxic Life in the Anthropocene" In *Anthropology for the Contemporary World: Exotic No More*, edited by Jeremy MacClancy, 223–40. Chicago: Chicago University Press.

Logan, Drake. 2011. "Toxic Violence: The Politics of Militarized Toxicity in Iraq and Afghanistan." *Annals of the Association of American Geographers*, 101(3): 253-83.

Longley, Hereward. 2013. "Razing Athabasca: Bitumen Extraction and the Industrial Colonization of North-Eastern Alberta, 1967–1983." Master's thesis, Memorial University of Newfoundland.

———. 2015. "Indigenous Battles for Environmental Protection and Economic Benefits during Commercialization of the Alberta Oil Sands, 1967–1986." In *Mining and Communities in Northern Canada: History, Politics, and Memory*, edited by. Arn Keeling and John Sandlos, 207–32. Calgary, AB: University of Calgary Press.

———. 2019a. "Conflicting Interests: Development Politics and the Environmental Regulation of the Alberta Oil Sands Industry, 1970–1980." *Environment and History*, 27(1): 97–125.

———. 2019b. "What Caused the Environmental Impacts of the Oil Sands Industry?" *White Horse Press* (blog), February 25. https://whitehorsepress.blog/2019/02/25/what-caused-the-environmental-impacts-of-the-oil-sands-industry/.

———. 2020. "Uncertain Sovereignty: Treaty 8, Bitumen, and Land Claims in the Athabasca Oil Sands Region." In *Extracting Home in the Oil Sands: Settler Colonialism and Environmental Change in Subarctic Canada*, edited by Clinton Westman, Tara Joly, and Lena Gross, 23–47. London: Routledge.

Lowe, Celia. 2006. *Wild Profusion: Biodiversity Conservation in an Indonesian Archipelago*. Princeton, NJ: Princeton University Press.

Lugo, Ariel. 1980. "Mangrove Issue Debates in Courtrooms." In *Proceedings, U.S. Fish and Wildlife Service Workshop on Coastal Ecosystems of the Southeastern United States*, edited by Robert Carey, Paul Markovits, and James Kirkwood, 48–60. Washington, DC: U.S. Fish and Wildlife Service, Office of Biological Services.

———. 2002. "Conserving Latin American and Caribbean Mangroves: Issues and Challenges." *Madera y Bosques* 8(1): 5–25.

Lugo, Ariel, and G. Cintrón. 1975. "The Mangrove Forests of Puerto Rico and Their Management." In *Proceedings of the International Symposium* on *Biology and Management of Mangroves*, edited by Gerald E. Walsh, Samuel C. Snedaker, and Howard J. Teas, 170–212. Gainesville: Institute of Food Agricultural Sciences, University of Florida.

Lugo, Ariel, and Samuel Snedaker. 1974. "The Ecology of Mangroves." *Annual Review of Ecology and Systematics* 5:39–64.

Lukács, Georg. 1971. *History and Class Consciousness: Studies in Marxist Dialectics.* Translated by R. Livingston. Cambridge, MA: MIT Press.

Lutz, Catherine, and Andrea Mazzarino, eds. 2019. *War and Health: The Medical Consequences of the Wars in Iraq and Afghanistan.* New York: New York University Press.

Lyons, Kristina. 2020. *Vital Decomposition: Soil Practitioners and Life Politics.* Durham, NC: Duke University Press.

Mabus, Ray. 2010. *America's Gulf Coast: A Long Term Recovery Plan after the Deepwater Horizon Oil Spill.* Washington, DC: EPA.

MacKenzie, Donald. 2008. *An Engine, Not a Camera: How Financial Models Shape Markets.* Cambridge, MA: MIT Press.

MacKenzie, Donald, Fabian Muniesa, and Lucia Siu, eds. 2007. *Do Economists Make Markets? On the Performativity of Economics.* Princeton, NJ: Princeton University Press.

MacLeish, Kenneth, and Zoe Wool. 2018. "US Military Burn Pits and the Politics of Health." *Medical Anthropological Quarterly*, Critical Care Series Online, August 1.

Mahony, James. 2016. "Fort McKay Chief Jim Boucher Believes You Can Have Oil Sands Pipelines and a Healthy Environment." *JW Energy News,* November 24.

Mahrane, Yannick, Marianna Fenzi, Céline Pessis, and Christophe Bonneuil. 2012. "From Nature to Biosphere: The Political Invention of the Global Environment." *Vingtième Siècle: Revue d'Histoire* 113(1): 127–41.

Malm, Andreas. 2016. *Fossil Capital: The Rise of Steam Power and the Roots of Global Warming.* New York: Verso.

———. 2018. *The Progress of This Storm: Nature and Society in a Warming World.* New York: Verso.

Marder, Michael. 2019. "Being Dumped." *Environmental Humanities* 11(1): 180–93.

Marino, Elizabeth. 2015. *Fierce Climate, Sacred Ground: Ethnography of Climate Change in Shishmaref, Alaska.* Fairbanks: University of Alaska Press.

Markowitz, Gerald, and David Rosner. 2002. *Deceit and Denial: The Deadly Politics of Industrial Pollution.* Berkeley: California University Press.

Martin, Laura. 2022. *Wild By Design: The Rise of Ecological Restoration.* Cambridge: Harvard University Press.

Martinez-Alier, Joan. 2002. *The Environmentalism of the Poor: A Study of Ecological Conflicts and Valuation.* Cheltenham, UK: Edward Elgar.

Marx, Karl. (1867) 1976. *Capital.* New York: Penguin.

Masco, Joseph. 2006. *The Nuclear Borderlands: The Manhattan Project in Post-Cold War New Mexico.* Princeton, NJ: Princeton University Press.

———. 2008. "Survival Is Your Business: Engineering Ruins and Affect in Nuclear America." *Cultural Anthropology* 23(2): 361–98.

———. 2010. "Bad Weather: On Planetary Crisis." *Social Studies of Science* 40(1): 7–40.

———. 2015. "The Age of Fallout." *History of the Present* 5(2): 137–68.

———. 2016. "Terraforming Planet Earth." In *The Politics of Globality since 1945: Assembling the Planet*, edited by Rens van Munster and Casper Sylvest, 44–70. London: Routledge.

———. 2021. *The Future of Fallout: And Other Episodes in Radioactive World-Making*. Durham, NC: Duke University Press.

Mathews, Andrew. 2010. *Instituting Nature: Authority, Expertise, and Power in Mexican Forests*. Cambridge, MA: MIT Press.

Mauer, Bill. 2004. "Ungrounding Knowledges Offshore: Caribbean Studies, Disciplinarity, and Critique." *Comparative American Studies* 2(3): 324–41.

McCarthy, John. 2018. "United Nations Report: U.S. Virgin Islands is the New Murder Capital of the Caribbean and the Fourth Overall in the World." *Virgin Islands Free Press*, January 23.

McDermott, Vince. 2014a. "Two More First Nations Drop Out of 'Unclear' Monitoring Program," *Fort McMurray Today*, January 23.

———. 2014b. "Last Local First Nation Leaves JOSM." *Fort McMurray Today*, May 2.

McIntosh, Robert. 1987. *The Background of Ecology: Concept and Theory*. Cambridge: Cambridge University Press.

McLachlan, Stéphane. 2014. "Water Is a Living Thing: Environmental and Human Health Implications of the Athabasca Oil Sands for the Mikisew Cree First Nation and Athabasca Chipewyan First Nation in Northern Alberta." Phase Two Report, submitted to Mikisew Cree First Nation, Athabasca Chipewyan First Nation, and University of Manitoba, July 7.

McNeill, J. R. 2000. *Something New under the Sun: An Environmental History of the Twentieth-Century World*. New York: Norton.

McNeill, J. R., and Peter Engelke. 2016. *The Great Acceleration*. Cambridge, MA: Harvard University Press.

McNeill, J. R., and Corinna Unger, eds. 2010. *Environmental Histories of the Cold War*. Cambridge, MA: Cambridge University Press.

Mech, Michelle. 2011. "A Comprehensive Guide to the Alberta Oil Sands." Report submitted to the Green Party of Canada, May.

Middle East Research and Information Project. 1974. "A Political Evaluation of the Arab Oil Embargo." *MERIP Reports* 23 (May): 23–25.

Miller, Pete, and Nikolas Rose. 1990. "Governing Economic Life." *Economy and Society* 19(1): 1–31.

Miller, Richard. 1979. *The Economy of the Virgin Islands*. Washington, DC: Office of Territorial Affairs, Department of Interior.

Miller, Shawn. 2003. "Stilt-Root Subsistence: Colonial Mangroves and Brazil's Landless Poor." *Hispanic American Historical Review* 83(2): 223–53.

———. 2007. *An Environmental History of Latin America*. Cambridge: Cambridge University Press.

Mills, C. Wright. 1959. *The Sociological Imagination*. Oxford: Oxford University Press.

Mintz, Sidney. 1966. "The Caribbean as a Socio-Cultural Area." *Cahiers d'histoire mondiale* 9(2): 912–37.

———. 1975. "The Caribbean Region." In *Slavery, Colonialism, and Racism*, edited by Sidney Mintz, 45–71. New York: Norton.

———. 1985. *Sweetness and Power*. New York: Penguin.

Mintz, Sidney, and Sally Price, eds. 1985. *Caribbean Contours*. Baltimore, MD: Johns Hopkins University Press.

Misrach, Richard, and Kate Orff. 2014. *Petrochemical America*. New York: Apature.

Mitchell, Timothy. 1998. "Fixing the Economy." *Cultural Studies* 12(1): 82–101.

———. 2011. *Carbon Democracy*. London: Verso.

Molotch, Harvey. 1970. "Oil in Santa Barbara and Power in America." *Sociological Inquiry* 40(1): 131–44.

Moore, Jason. 2015. *Capitalism in the Web of Life: Ecology and the Accumulation of Capital*. New York: Verso.

Moore, Jason W., and Raj Patel. 2018. *A History of the World in Seven Cheap Things*. Berkeley: University of California Press.

Moran-Thomas, Amy. 2019. *Traveling with Sugar: Chronicles of a Global Epidemic*. Berkeley: University of California Press.

Morimoto, Ryo. 2022. "A Wild Boar Chase: Ecology of Harm and Half-Life Politics in Coastal Fukushima." *Cultural Anthropology* 37(1): 69–98.

Morse, Kathryn. 2012. "There Will Be Birds: Images of Oil Disasters in the Nineteenth and Twentieth Centuries." *Journal of American History* 99(1): 124–34.

Mosk, Stanley. 1976. "William O. Douglas." *Ecology Law Quarterly* 5(2): 229–32.

Mumford, Lewis. 1963. *The Highway and the City*. New York: Harcourt.

Mumme, Stephen, C. Richard Bath, and Valerie Assetto. 1988. "Political Development and Environmental Policy in Mexico." *Latin American Research Review* 23(1): 7–34.

Murphy, Michelle. 2006. *Sick Building Syndrome and the Problem of Uncertainty: Environmental Politics, Technoscience, and Women Workers*. Durham, NC: Duke University Press.

———. 2008. "Chemical Regimes of Living." *Environmental History* 13(4): 695–703.

———. 2017. "Alterlife and Decolonial Chemical Relations." *Cultural Anthropology* 32(4): 494–503.

Nadasdy, Paul. 1999. "The Politics of TEK: Power and the Integration of Knowledge." *Arctic Anthropology* 36(1/2): 1–18.

———. 2005. "The Anti-Politics of TEK: The Institutionalization of Co-Management Discourse Practice." *Anthropologica* 47(2): 215–32.

Nadeau, Royal J., and Eugene T. Bergquist. 1977. "Effects of the March 18, 1973 Oil Spill Near Cabo Rojo, Puerto Rico on Tropical Marine Communities." *Proceedings of International Oil Spill Conference* 1:535–38.

Nading, Alex. 2020. "Living in a Toxic World." *Annual Review of Anthropology* 49:209–24.

Nash, Roderick. 1990. *Wilderness and the American Mind*. New Haven, CT: Yale University Press.

National Academy of Sciences. 1969. *Effects of Chronic Exposure to Low Levels of Carbon Monoxide*. Washington, DC: National Academy of Sciences.

National Oceanic and Atmospheric Administration (NOAA). 2010. "BP Deepwater Horizon Oil Budget: What Happened to the Oil?" NOAA report, August 30.

National Research Council. 1975. *Petroleum in the Marine Environment*. Washington, DC: National Academy Press.

———. 1985. *Oil in the Sea: Inputs, Fates, and Effects*. Washington, DC: National Academy Press.

Navaro, Yael. 2020. "The Aftermath of Mass Violence: A Negative Methodology," *Annual Review of Anthropology* 49(October): 161–73.

Netting, Robert. 1977. *Cultural Ecology*. Menlo Park, CA: Benjamin/Cummings.

New Left Review. 1982. Introduction to *Exterminism and Cold War*. New York: Verso.

New York State Department of Health. 2017. "Cancer Incidence Investigation, 1995–2014: Village of Hoosick Falls." Albany, NY.

Newcombe, H. B. 1957. "Magnitude of Biological Hazard from Strontium-90." *Science* 126(3273): 549–51.

Nicholson, Max. 1970. *The Environmental Revolution: A Guide for the New Masters of the Earth*. New York: McGraw-Hill.

Nikiforuk, Andrew. 2010. "A Smoking Gun on Athabasca River: Deformed Fish." *The Tyee*, September 17.

Nixon, Richard. 1973. "Address to the Nation about Policies to Deal With Energy Shortages." November 7.

Nixon, Rob. 2011. *Slow Violence and the Environmentalism of the Poor*. Cambridge, MA: Harvard University Press.

Norris, Robert, and Hans Kristensen. 2010. "Global Nuclear Weapons Inventories, 1945–2010." *Bulletin of the Atomic Scientists* 66(4): 77–83.

Norton, Peter. 2011. *Fighting Traffic: The Dawn of the Motor Age in the American City*. Cambridge, MA: MIT Press.

Novy, Andreas. 2020. "The Political Trilemma of Contemporary Social-Ecological Transformation: Lessons from Karl Polanyi's The Great Transformation." *Globalizations* December 4.

Nuttall, W., C. Samaras, and M. Brazilian. 2017. "Energy and the Military: Convergence of Security, Economic, and Environmental Decision-Making." University of Cambridge Energy Policy Group Report 1717, 1–29. Cambridge.

Nuwer, Rachel. 2013. "Oil Sands Mining Uses Up Almost as Much Energy as It Produces." *Inside Climate News*, February 19.

O'Connor, James. 1991. *Natural Causes: Essays in Ecological Marxism*. New York: Guilford Press.

———. 1998. *Natural Causes: Essays in Ecological Marxism*. New York: Guilford Press.

———. 2009. "On the Two Contradictions of Capital." *Capitalism, Nature, Socialism* 2(3): 107–9.

Odum, Eugene. 1953. *Fundamentals of Ecology*. Philadelphia, PA: W. B. Saunders.

Odum, William. 1970. "Pathways of Energy Flow in a South Florida Estuary." PhD diss., University of Miami.

Odum, William, and R. E. Johannes. 1975. "The Response of Mangroves to Man-Induced Environmental Stress." In *Tropical Marine Pollution*, edited by E. J. Furguson Wood and R. E. Johannes, 52–62. Elsevier Oceanography Series. Amsterdam: Elsevier Publishing.

Odum, William E., Carole C. McIvor, and Thomas J. Smith III. 1982. *The Ecology of the Mangroves of South Florida: A Community Profile*. Washington, DC: US Department of the Interior, Bureau of Land Management, Fish and Wildlife Service.

Ogden, Laura. 2011. *Swamplife: People, Gators, and Mangroves Entangled in the Everglades*. Minneapolis: University of Minnesota Press.

Oldenziel, Ruth. 2011. "Islands: The United States as Networked Empire." In *Entangled Geographies: Empire and Technopolitics in the Global Cold War*, edited by Gabrielle Hecht, 13–42. Cambridge, MA: MIT Press.

Ong, Aihwa. 1987. *Spirits of Resistance and Capitalist Discipline: Factory Women in Malaysia*. Albany: State University of New York Press.

Ottinger, Gwen. 2013. *Refining Expertise: How Responsible Engineers Subvert Environmental Justice Challenges*. New York: New York University Press.

Paget, Henry. 1985. *Peripheral Capitalism and Underdevelopment in Antigua*. London: Transaction Publishers.

Paiewonsky, Ralph M., with Isaac Dookhan. 1990. *Memoirs of a Governor: A Man for the People*. New York: New York University Press.

Pantojas-Garcia, Emilio. 1990. *Development Strategies as Ideology: Puerto Rico's Export-Led Industrialization Experience*. Boulder: Lynne Rienner.

Parreñas, Juno Salazar. 2018. *Decolonizing Extinction: The Work of Orangutan Rehabilitation*. Durham, NC: Duke University Press.

Pasternak, Judy. 2011. *Yellow Dirt: A Poisoned Land and the Betrayal of the Navajos*. New York: Free Press.

Payne, Anthony, and Paul Sutton, eds. 1984. *Dependency under Challenge: The Political Economy of the Commonwealth Caribbean*. Manchester, UK: Manchester University Press.

Peluso, Nancy. 1992. *Rich Forests, Poor People: Resource Control and Resistance in Java*. Berkeley: University of California Press.

Perrow, Charles. 1984. *Normal Accidents: Living with High-Risk Technology*. Princeton, NJ: Princeton University Press.

Petryna, Adriana. 2002. *Life Exposed: Biological Citizens after Chernobyl*. Princeton, NJ: Princeton University Press.

———. 2015. "What Is a Horizon? Navigating Thresholds in Climate Change Uncertainty." In *Modes of Uncertainty*, edited by L. Samimian-Darash and P. Rabinow, 147–64. Chicago: University of Chicago Press.

———. 2022. *Horizon Work: At the Edges of Knowledge in an Age of Runaway Climate Change*. Princeton, NJ: University of Princeton Press.

Pinch, Trevor, and Wiebe Bijker. 1984. "The Social Construction of Facts and Artifacts: Or How the Sociology of Science and the Sociology of Technology Might Benefit Each Other." *Social Studies of Science* 14:399–441.

Pobicki, James. 1980. "U.S. Virgin Islands Economy: A Review of the 1970s and Outlook for the 1980s." *Annual Economic Review*. Virgin Island Archive Room, Florence Williams Public Library, Christiansted, St. Croix Office of Policy, Planning, and Research, Department of Commerce.

Pollock, Anne. 2016. "Queering Endocrine Disruption." In *Object-Oriented Feminism*, edited by Katherine Behar, 183–99. Minneapolis: Minnesota University Press.

Porter, Theodore. 1995. *Trust in Numbers: The Pursuit of Objectivity in Science and Public Life*. Princeton, NJ: Princeton University Press.

Poujade, Robert. 1975. *Le ministère de l'impossible*. Paris: Calmann-Lévy.

Povinelli, Elizabeth. 2017. "Fires, Fogs, Winds." *Journal of Cultural Anthropology* 32(4): 504–13.

Powell, David. 1973. *Problems of Economic Development in the Caribbean*. Washington, DC: British/North America Committee.

Powers, Michael. 2004. *The Risk Management of Everything: Rethinking the Politics of Uncertainty*. London: Demos.

Price, Richard, and Sally Price. 1997. "Shadowboxing in the Mangrove." *Cultural Anthropology* 12(1): 3–36.

Rabinow, Paul. 1996. *Making PCR: A Story of Biotechnology*. Chicago: University of Chicago Press.

Radkau, Joachim. 2014. *The Age of Ecology*. Cambridge, UK: Polity.

Raffles, Hugh. 2002. *In Amazonia: A Natural History*. Princeton, NJ: Princeton University Press.

Raines, Ben. 2010. "BP Buys up Gulf Scientists for Legal Defense, Roiling Academic Community." *Mobile Press Register*, July 16.

Rajak, Dinah. 2011. *In Good Company: An Anatomy of Corporate Social Responsibility*. Stanford, CA: Stanford University Press.

Ramsaran, Ramesh. 1989. *The Commonwealth Caribbean in the World Economy*. London: Macmillan.

Ranciére, Jacques. 2001. "Ten Thesis on Politics." *Theory and Event* 5(3): 2–7.

Rappaport, Roy. 1968. *Pigs for the Ancestors: Ritual in the Ecology of a New Guinea People*. New Haven, CT: Yale University Press.

Reisman, David. 1964. *Abundance for What?* Garden City, NJ: Doubleday.

Reiss, Louise Zibold. 1961. "Strontium-90 Absorption by Deciduous Teeth." *Science* 134(3491): 1669–73.

Rich, Nathaniel. 2016. "The Lawyer Who Became DuPont's Worst Nightmare." *New York Times Magazine*, January 6, 20–28.

Richardson, Bonham. 1992. *The Caribbean in the Wider World, 1492–1992: A Regional Geography*. Cambridge: Cambridge University Press.

Rienow, Robert, and Leona Rienow. 1967. "The Oil around Us." *New York Times Magazine*, June 4.

Robbins, Paul, Nik Heynen, James McCarthy, and Scott Prudham, eds. 2007. *Neoliberal Environments: False Promises and Unnatural Consequences*. London: Routledge.

Roberts, Celia. 2003. "Drowning in a Sea of Estrogens: Sex Hormones, Sexual Reproduction and Sex." *Sexualities* 6(2): 195–213.

Roberts, Elizabeth. 2017. "What Gets Inside: Violent Entanglements and Toxic Boundaries in Mexico City." *Cultural Anthropology* 32(4): 592–619.

Robertson, Thomas. 2008. "This Is the American Earth: American Empire, the Cold War, and American Environmentalism." *Diplomatic History* 32(4): 561–84.

Rockström, Johan, Owen Gaffney, Joeri Rogelj, Malte Meinshausen, Nebojsa Nakicenovic, and Hans Joachim Schellnhuber. 2017. "A Roadmap for Rapid Decarbonization." *Science* 355(6331): 1269–71.

Rockström, Johan, Will Steffen, Kevin Noone, Åsa Persson, F. Stuart Chapin III, Eric F. Lambin, Timothy M. Lenton, Marten Scheffer, Carl Folke, Hans Joachim Schellnhuber, Björn Nykvist, Cynthia A. de Wit, Terry Hughes, Sander van der Leeuw, Henning Rodhe, Sverker Sörlin, Peter K. Snyder, Robert Costanza, Uno Svedin, Malin Falkenmark, Louise Karlberg, Robert W. Corell, Victoria J. Fabry, James Hansen, Brian Walker, Diana Liverman, Katherine Richardson, Paul Crutzen, and Jonathan A. Foley. 2009. "Planetary Boundaries: Exploring the Safe Operating Space for Humanity." *Ecology and Society* 14(2): 32–45.

Rogers, Doug. 2015. *The Depths of Russia: Oil, Power, and Culture after Socialism*. Ithaca, NY: Cornell University Press.

Roitman, Janet. 2013. *Anti-Crisis*. Durham, NC: Duke University Press.

Rome, Adam. 2013. *The Genius of Earth Day*. New York: Macmillan.

Root, Al. 2019. "Sizing Up an Environmental Liability for 3M and Others." *Barrons*, August 1.

Rose, Nikolas. 2006. *The Politics of Life Itself: Biomedicine, Power, and Subjectivity*. Princeton, NJ: Princeton University Press.

Rosenbaum, Walter. 1977. *The Politics of Environmental Concern*. New York: Praeger.

Rosner, David, and Gerald Markowitz, eds. 1989. *Dying for Work: Workers' Safety and Health in Twentieth Century America*. New York: Midland Books.

Rubaii, Kali. 2020. "Birth Defects and the Toxic Legacy of War in Iraq." *Middle East Research and Information Project* 296 (Fall).

Rudd, Robert L. 1964. "Pesticides and the Living Landscape." *Soil Science* 98(2): 144–45.

Ruddiman, William. 2005. *Plows, Plagues, and Petroleum: How Humans Took Control of Climate*. Princeton, NJ: Princeton University Press.

Russel, Edmund. 2001. *War and Nature: Fighting Humans and Insects with Chemicals from World War I to Silent Spring*. Cambridge: Cambridge University Press.

Rützler, Klaus, and Wolfgang Sterrer. 1970. "Oil Pollution: Damage Observed in Tropical Communities along the Atlantic Seaboard of Panama." *BioScience* 20(4): 222–24.

Sagan, Carl. 1983. "Nuclear Winter." *Parade Magazine*, February 3.

Samimian-Darash, Limor. 2009. "A Pre-Event Configuration for Biological Threats: Preparedness and the Constitution of Biosecurity Events." *American Ethnologist* 36(3): 478–91.

Sawyer, Suzana. 2004. *Crude Chronicles: Indigenous Politics, Multinational Oil, and Neoliberalism in Ecuador*. Durham, NC: Duke University Press.

Sayers, R., F. Meriwhether, and W. Yant. 1922. "Physiological Effects of Exposure to Low Concentrations of Carbon Monoxide." *Public Health Reports*, no. 37.

Schereschewsky, J. W. 1915. "Industrial Hygiene: A Plan for Education in the Avoidance of Occupational Diseases and Injuries." *Public Health Reports*, no. 30: 2934–35.

Schlosser, Eric. 2013. *Command and Control: Nuclear Weapons, the Damascus Accident, and the Illusion of Safety*. New York: Penguin.

Schnaiberg, Allen. 1980. *The Environment: From Surplus to Scarcity*. Oxford: Oxford University Press.

Scott, James. 1985. *Weapons of the Weak: Everyday Forms of Peasant Resistance*. New Haven, CT: Yale University Press.

Scott, Rebecca. 2010. *Removing Mountains: Extracting Nature and Identity in the Appalachian Coal Fields*. Minneapolis: Minnesota University Press.

Sellers, Christopher. 1994. "Factory as Environment: Industrial Hygiene, Professional Collaborations, and the Modern Sciences of Pollution." *Environmental History Review* 18(1): 55–83.

———. 1997. *Hazards of the Job: From Industrial Disease to Environmental Health Science*. Chapel Hill: University of North Carolina Press.

———. 1999. "Body, Place and the State: The Makings of an 'Environmentalist' Imaginary in the Post-World War II U.S." *Radical History Review* no. 74: 31–64.

———. 2012. *Crabgrass Crucible: Suburban Nature and the Rise of Environmentalism in Twentieth Century America*. Chapel Hill: University of North Carolina Press.

Shapin, Steven. 1994. *The Social History of Truth: Civility and Science in Seventeenth-Century England*. Chicago: University of Chicago Press.

Shapiro, Nicholas. 2015. "Attuning to the Chemosphere: Domestic Formaldehyde, Bodily Reasoning, and the Chemical Sublime." *Cultural Anthropology* 30(3): 368–93.

Shaw, Rosalind. 1992. "'Nature,' 'Culture' and Disasters: Floods and Gender in Bangladesh." In *Bush Base, Forest Farm: Culture, Environment, and Development*, edited by Elisabeth Croll and David Parkin, 202–17. London: Routledge.

Sheller, Mimi. 2003. *Consuming the Caribbean: From Arawaks to Zombies*. London: Routledge.

———. 2019. "The Origins of Global Carbon Form." *Log* 47(Fall): 56–68.

———. 2020. *Island Futures: Caribbean Survival in the Anthropocene*. Durham, NC: Duke University Press.

Shepard, Paul. 1958. "The Place of Nature in Man's World," *School Science and Mathematics*, 58(May): 394–403.

———. 1969. *The Subversive Science: Essays Towards an Ecology of Man*. Boston: Houghton Mifflin.

———. 1970. "The Environment." *New York Times Book Review*, August 30.

Shever, Elana. 2012. *Resources for Reform: Oil and Neoliberalism in Argentina*. Stanford, CA: Stanford University Press.

Shiva, Vandana. 1989. *The Violence of the Green Revolution*. London: Zed Books.

Sieferle, Rolf Peter. (1982) 2001. *The Subterranean Forest: Energy Systems and the Industrial Revolution*. Translated by Michael Osman. Cambridge: Cambridge University Press.

Simpson, Audra. 2014. *Mohawk Interruptus: Political Life across the Borders of Settler States*. Durham, NC: Duke University Press.

Singer, Merrill. 2011. "Down Cancer Alley: The Lived Experience and Environmental Suffering in Louisiana's Chemical Corridor." *Medical Anthropology Quarterly* 25(2): 141–63.

Siuzdak, Gary. 2004. "An Introduction to Mass Spectrometry." *Journal of the Association of Laboratory Automation* 9(2): 50–63.

Sloterdijk, Peter. 2009. *Terror from the Air*. Translated by Amy Patton and Steve Corcoran. Los Angeles: Semiotext(e).

Snedaker, Samuel C., Patrick D. Biber, and Rafael J. Aravjo. 1997. "Oil Spills and Mangroves: An Overview." In *Managing Oil Spills in Mangrove Ecosystems: Effects, Remediation, Restoration, and Modeling*, edited by C. Edward Proffitt, 1–18. OCS Study MMS 97-0003. New Orleans: US Department of the Interior, Minerals Management Service, Gulf of Mexico OCS Region, New Orleans.

Smargiassi, Audrey, Ian Kosatsky, John Hicks, Céline Plante, Ben Armstrong, Paul J. Villeneuve, and Sophie Goudreau. 2009. "Risk of Asthmatic Episodes in Children Exposed to Sulfur Dioxide Stack Emissions from a Refinery Point Source in Montreal, Canada." *Environmental Health Perspectives* 117(4): 653–59.

Smith, Linda Tuhiwai. 1999. *Decolonizing Methodologies: Research and Indigenous Peoples*. London: Zed Books.

Smith, Zadie. 2014. "Elegy for a Country's Seasons." *New York Review of Books*, April 3, 3–7.

Spear, Wayne. 2013. "An Interview with Dave Tuccaro." *Wayne Spear Reports*, January 4. https://waynekspear.com/2013/01/04/dave-tuccaro/.

Stamatopoulou-Robbins, Sophia. 2019. *Waste Siege: The Life of Infrastructure in Palestine*. Stanford, CA: Stanford University Press.

Steinbeck, John. 1951. *The Log from the Sea of Cortez*. New York: Penguin.

Steward, Julian. 1955. "The Concept and Method of Cultural Ecology." In *Theory of Cultural Change: The Methodology of Multilinear Evolution*, edited by 30–42. Urbana: University of Illinois Press.

Stoetzer, Bettina. 2018. "Ruderal Ecologies: Rethinking Nature, Migration, and the Urban Landscape in Berlin." *Cultural Anthropology* 33(2): 295–323.

Stoler, Ann, ed. 2013. *Imperial Debris: On Ruins and Ruination*. Durham, NC: Duke University Press.

Stoler, Ann Laura. 1995. *Capitalism and Confrontation in Sumatra's Plantation Belt, 1870–1979*. Ann Arbor: University of Michigan Press.

———. 2008. "Epistemic Politics: Ontologies of Colonial Common Sense." *Philosophical Forum* 39(3): 349–61.

Stoler, Ann Laura, and Carole McGranahan. 2007. "Introduction: Refiguring Imperial Terrain." In *Imperial Formations*, edited by Ann Laura Stoler, Carole McGranahan, and Peter Perdue, 3–4. Santa Fe: School of American Research Press.

Stone, Christopher D. 1972. "Should Trees Have Standing? Toward Legal Rights for Natural Objects." *Southern California Law Review* 45(2): 450–501.

Sweezy, Paul. 1972. "Cars and Cities." *Monthly Review* 23(11): 1–18.

Tarr, Joel A. 1996. *The Search for the Ultimate Sink: Urban Pollution in Historical Perspective*. Akron, OH: University of Akron Press.

Taussig, Michael. 1980. *The Devil and Commodity Fetishism in South America*. Chapel Hill: University of North Carolina Press.

———. 1991. *Shamanism, Colonialism, and the Wild Man: A Study in Terror and Healing*. Chicago: University of Chicago Press.

———. 2000. "The Beach (A Fantasy)." *Critical Inquiry* 26(2): 248–78.

———. 2018. *Palma Africana*. Chicago: University of Chicago Press.

Taves, Donald. 1968. "Evidence That There Are Two Forms of Fluoride in Human Serum." Nature 217(5133): 1050–51.

Taylor, Bron, ed. 1995. *Ecological Resistance Movements: The Global Emergence of Radical Environmentalism*. Albany: State University of New York Press.

Taylor, Dorceta 2014. *Toxic Communities*. New York: New York University Press.

Terkel, Studs. 1979. "Politics of Energy: Interview with Barry Commoner." Radio Broadcast, June 18. Studs Terkel Radio Archive, The Chicago History Museum.

Theman, Trevor. 2009. "Investigation Report." Filed with College of Physicians and Surgeons of Alberta, November 4.

Therrien, James. 2017. "Teflon Town." *Vermont Digger*, September.

Thomä, Dieter. 2001. "The Name on the Edge of Language: A Complication in Heidegger's Introduction to Metaphysics." In *A Companion to Heidegger's Introduction to Metaphysics*, edited by Richard Polt and Gregory Fried, 103–22. New Haven, CT: Yale University Press.

Thomas, William, ed. 1956. *Man's Role in Changing the Face of the Earth*. With the collaboration of Carl Sauer, Marston Bates, and Lewis Mumford. Chicago: University of Chicago Press.

Thompson, E. P. 1971. "The Moral Economy of the English Crowd in the Eighteenth Century," *Past and* Present 50(February): 76–136.

———. 1982. "Notes on Exterminism, the Last Stage of Civilization." In *Exterminism and the Cold War*, edited by New Left Review, 1–34. London: Verso.

Thurland, Karen. 1979. *Ralph Paiewonsky: Economic and Social Reformer of the Virgin Islands*. Garden City, NY: Adelphi University Press.

Ticktim, Miriam. 2017. "Invasive Others: Towards a Contaminated World." *Social Research* 84(1): xxi–xxxiv.

Timoney, Kevin. 2007. "A Study of Water and Sediment Quality as Related to Public Health Issues, Fort Chipewyan, Alberta." Report Prepared for Nunee Health Board Society, Fort Chipewyan, Alberta, November.

———. 2013. *The Peace-Athabasca Delta: Portrait of a Dynamic Ecosystem*. Edmonton: University of Alberta Press.

Timoney, Kevin, and Peter Lee. 2009. "Does the Alberta Tar Sands Industry Pollute? The Scientific Evidence." *Open Conservation Biology Journal* 3:65–81.

———. 2011. "Polycyclic Aromatic Hydrocarbons Increase in Athabasca River Delta Sediment: Temporal Trends and Environmental Correlates." *Environmental Science and Technology* 45(10): 4278–84.

Todd, Zoe. 2017. "Fish, Kin, and Hope." *Afterall* 43 (Spring): 102–7.

Tomilson, P. B. 1986. *The Botany of Mangroves*. Cambridge: Cambridge University Press.

Torry, William. 1979. "Anthropological Studies in Hazardous Environments: Past Trends and New Horizons." *Current Anthropology* 20(3): 517–40.

Train, Russell E. 1978. "The Environment Today." *Science* 201(4353): 320–24.

Tri-Island Economic Development Council. 1983. "The Socio-Economic and the Environmental Impact of Energy Facilities in the U.S. Virgin Islands." Report prepared for US Department of Commerce. Virgin Island Archive Room, Florence Williams Public Library, Christiansted, St. Croix.

Trouillot, Michel-Rolph. 1992. "The Caribbean Region: An Open Frontier in Anthropological Theory." *Annual Review of Anthropology* 21:19–42.

Tsing, Anna, Heather Swanson, Elaine Gan, and Nils Bubandt. 2017. *Arts of Living on a Damaged Planet*. Minneapolis: Minnesota University Press.

Tsing, Anna Lowenhaupt. 2017. *The Mushroom at the End of the World: On the Possibility of Life in Capitalist Ruins*. Princeton, NJ: Princeton University Press.

Tuck, Eve. 2009. "Suspending Damage." *Harvard Educational Review* 79(3): 409–27.

Tyas, Michael. 2014. "Clear and Worrisome Fort Chipywan Health Report Going Pubic Monday." *One River News*, July 3.

Tyson, George. 1991. "The Homestead Program on St. Croix." Report for the St. Croix Historical Society, Christiansted, St. Croix.

United Church of Christ. 1987. "Toxic Waste and Race."

United Nations (UN). 1972. *Report of the United Nations Conference on the Human Environment*. Stockholm, June 5–16. New York: United Nations Publications.

———. 1973. *Report of the United Nations Conference on the Human Environment*. New York: United Nations Publications.

———. 1979. *Yearbook of World Energy Statistics*. Geneva: United Nations Publications.

———. 1980. *Yearbook of World Energy Statistics*. Geneva: United Nations Publications.

———. 2021. "COP26: Enough of 'Treating Nature Like a Toilet'—Guterres Brings Stark Call for Climate Action to Glasgow." UN news release, November 1. https://news.un.org/en/story/2021/11/1104542.

United States Coast Guard (USCG). 2010. "Sub-Sea and Sub-Surface Oil and Dispersant Detection, Sampling, and Monitoring Strategy." Unified Command Directive, August 18.

United States Virgin Islands (USVI) Source. 2013. "One Year Post-HOVENSA: Worse Before It Gets Better?" *USVI Source*, January 26.

Urry, John. 2009. *Societies Beyond Oil: Oil Dregs and Social Futures*. London: Zed Books.

———. 2013. *Societies Beyond Oil*. London: Zed Books.

Valentine, David, John Kessler, Molly Redmond, Stephanie Mendes, Monica Heintz, Christopher Farwell, Lei Hu, Franklin Kinnaman, Shari Yvon-Lewis, Mengran Du, Eric Chan, Fenix Tigreros, and Christie Villanueva. 2010. "Propane Respiration Jump-Starts Microbial Response to a Deep Oil Spill." *Science* 330(6001): 208–11.

Van Dooren, Thomas. 2016. *Flight Ways: Life and Loss at the Edge of Extinction*. New York: Columbia University Press.

Vanderklippe, Nathan. 2012. "In Oil Sands, A Native Millionaire Sees 'Economic Force' for First Nations." *Globe and Mail*, August 13.

Vaughan, Diane. 1997. *The Challenger Launch Decision: Risky Technology, Culture, and Deviance at NASA*. Chicago: University of Chicago Press.

Vaughn, Sarah. 2012. "Reconstructing the Citizen: Disaster, Citizenship, and Expertise in Racial Guyana." *Critique of Anthropology* 32(4): 359–87.

———. 2017. "Disappearing Mangroves: The Epistemic Politics of Climate Adaptation in Guyana." *Cultural Anthropology* 32(2): 242–68.

———. 2022. *Engineering Vulnerability: In Pursuit of Climate Adaptation*. Durham, NC: Duke University Press.

Vázquez-Arroyo, Antonio. 2008. "Universal History Disavowed: On Critical Theory and Postcolonialism." *Postcolonial Studies* 11(4): 451–73.

Veblen, Thorsten. 1899. *The Theory of the Leisure Class: An Economic Study in the Evolution of Institutions*. New York: MacMillan.

Virgin Islands Office of Public Relations and Information. 1967. *Your Virgin Islands*. Charlotte Amalie: Virgin Islands Press.

Vitalis, Robert. 2020. *Oilcraft: The Myths of Scarcity and Security that Haunt U.S. Energy Policy*. Stanford, CA: Stanford University Press.

Viveiros de Castro, Eduardo. 2014. *Cannibal Metaphysics*. Minneapolis: University of Minnesota Press.

Voyles, Tracy. 2015. *Wastelanding: Legacies of Uranium Mining in Navajo Country*. Minneapolis: Minnesota University Press.

Walcott, Derek. 1986. *Collected Poems, 1948–1984*. New York: Farrar, Strauss, and Giroux.

Walker, Mark, and Zolan Kanno-Youngs. 2019. "FEMA's Hurricaine Aid to Puerto Rico and the Virgin Islands Has Stalled." *New York Times*, November 27.

Walley, Christine. 2004. *Rough Waters: Nature and Development in an East African Marine Park*. Princeton, NJ: Princeton University Press.

Ward, Barbara, and Rene Dubos. 1972. *Only One Earth: The Care and Maintenance of a Small Planet*. New York: Norton.

Warde, Paul, Libby Robin, and Sverket Sörlin. 2018. *The Environment: A History of the Idea*. Baltimore, MD: Johns Hopkins University Press.

Watt-Cloutier, Sheila. 2015. *The Right to Be Cold: One Woman's Story of Protecting Her Culture, the Arctic, and the Whole Planet*. New York: Penguin.

Watts, Michael J. 2005. "Righteous Oil? Human Rights, the Oil Complex, and Corporate Social Responsibility." *Annual Review of Environment and Resources* 30:373–407.

Watts, Michael, and Nancy Peluso, eds. 2001. *Violent Environments*. Ithaca, NY: Cornell University Press

Wayland, Mark, John Headley, Kerry Peru, Robert Crosley, and Brian Brownlee. 2008. "Levels of Polycyclic Aromatic Hydrocarbons and Dibenzothiophenes in Wetland Sediments and Aquatic Insects in the Oil Sands Area of Northeastern Alberta, Canada." *Environmental Monitoring Assessments* 136(1–3): 167–82.

Weber, Bob. 2014. "Higher Cancer Rates Not Found in Oilsands Community, Study Shows." CBC, March 24. www.cbc.ca/news/canada/edmonton/higher-cancer-rates-not-found-in-oilsands-community-study-shows-1.2584323#:~:text=CBC%20News%20Loaded-,Higher%20cancer%20rates%20not%20found%20in%20oilsands%20community%2C%20study%20shows,have%20higher%20overall%20cancer%20rates.

Weber, Max. (1920) 2002. *The Protestant Ethic and the Spirit of Capitalism.* Translated by Stephen Kalberg. New York: Routledge.

Weinhold, Bob, and Susan Booker. 2011. "Oil Sands Contaminants: Editors Response." *Environmental Health Perspectives* 119(8): A330–31.

Weiss, Margaret. 2016. "Always After: Desiring Queerness, Desiring Anthropology." *Cultural Anthropology* 31(4): 627–38.

Weizman, Eyal. 2007. *Hollow Land: Israel's Architecture of Occupation.* New York: Verso.

Welker, Marina. 2009. "'Corporate Security Begins in the Community': Mining, the Corporate Social Responsibility Industry, and Environmental Advocacy in Indonesia." *Cultural Anthropology* 24(1): 142–79.

———. 2014. *Enacting the Corporation.* Berkeley: California University Press.

Wells, Christopher. 2012. *Car Country: An Environmental History.* Seattle: University of Washington Press.

West, Paige. 2006. *Conservation Is Our Government Now: The Politics of Ecology in Papua New Guinea.* Durham, NC: Duke University Press.

Weszkalnys, Gisa. 2014. "Anticipating Oil: The Temporal Politics of a Disaster Yet to Come." *Sociological Review* 62(S1): 211–35.

Whalen, Carmen. 2001. *From Puerto Rico to Philadelphia: Puerto Rico Workers and Postwar Economies.* Philadelphia: Temple University Press.

Whitaker, John. 1988. "Earth Day Recollections: What it Was Like When the Movement Took Off." *EPA Journal* (July/August): 14–18.

White, Raymond. 1964. "Pesticides." *Science* 145(3633): 729–30.

White House. 1965. "Restoring the Quality of Our Environment." Report of the Environmental Pollution Panel, President's Science Advisory Committee, November. Washington, DC: US Government Printing Office.

Why Files. 2008. "Tar Sands: Heavy Price for Heavy Oil?" October 2009. https://whyfiles.org/314tarsands/index.php?g=3.txt.

Wilcox, Fred. 2011. *Scorched Earth: Legacies of Chemical Warfare in Vietnam.* Boston: Seven Stories.

Wilder, Gary. 2012. "Anticipation." *Political Concepts: A Lexicon,* May 24. www.politicalconcepts.org/anticipation-gary-wilder/.

Williams, Eric. 1944. *Capitalism and Slavery.* Chapel Hill: University of North Carolina Press.

———. 1970. *From Columbus to Castro: The History of the Caribbean, 1492–1969.* London: Deutsch.

Williams, Raymond. 1973. *The Country and the City.* Oxford: Oxford University Press.

———. 1976. *Keywords: A Vocabulary of Culture and Society.* Oxford: Oxford University Press.

———. 1977. *Marxism and Literature.* Oxford: Oxford University Press.

———. 1979. *Politics and Letters: Interviews with New Left Reviews.* London: Verso.

Wolf, Eric R. 1965. *Anthropology.* New York: Prentice Hall.

———. 1969. *Peasant Wars of the Twentieth Century.* New York Harper and Row.

———. 1982. *Europe and the People without History.* Berkeley: University of California Press.

Woodhouse, Edward. 2011. "Conceptualizing Disaster as Extreme Versions of Everyday Life." In *Dynamics of Disaster: Lessons on Risk, Response, and Recovery*, edited by Rachel Dowty and Barbara Allen, 61–76. Washington, DC: Earthscan Press.

Woodwell, George. 1967. "Toxic Substances and Ecological Cycles." *Scientific American* 216(3): 24–31.

———. 1970. "Effects of Pollution on the Structure and Physiology of Ecosystems." *Science* 168(3930): 429–33.

Wool, Zoë. 2017. "The Relativity of Toxicity." Anthrodendum, December 8. https://anthrodendum.org/2017/12/08/the-relativity-of-toxicity/.

World Bank. 1984. "World Refinery Industry: Need for Restructuring." World Bank Technical Paper 32. Washington, DC: World Bank.

———. 2017. "Electricity Production, Sources, and Access." In *World Development Indicators*. Washington, DC: World Bank.

World Health Organization (WHO). 1983. "Effects of Nuclear War on Health and Health Services." Report to the Thirty-Sixth World Health Assembly, Geneva, May 13.

Worster, Donald. 1994. *Nature's Economy: A History of Ecological Ideas.* Cambridge: Cambridge University Press.

Wylie, Sara. 2018. *Fractivism: Corporate Bodies and Chemical Bonds.* Durham, NC: Duke University Press.

Yelvington, Kevin. 1995. *Producing Power: Ethnicity, Gender, and Class in a Caribbean Workplace.* Philadelphia: Temple University Press.

Yost, Nicholas. 1992. "Rio and the Road Beyond." *Environmental Law* 11(4): 1–6.

Zhu, Wenyu, Harrison Roakes, Stephen Zembra, and Appala Raju Badireddy. 2019. "PFAS Background in Vermont Shallow Soils." Report commissioned by Vermont Department of Environmental Conservation (DEC). March 24. https://anrweb.vt.gov/PubDocs/DEC/PFOA/Soil-Background/PFAS-Background-Vermont-Shallow-Soils-03-24-19.pdf.

Zuev, Dennis. 2018. "Digital Afterlife: (Eco)civilizational Politics of the Site and the Sight of E-waste in China." *Anthropology Today* 34(6): 11–15.

Zwick, David. 1971. *Water Wasteland: Ralph Nader's Study Group Report on Water Pollution.* New York: Grossman.

Index

Founded in 1893,
UNIVERSITY OF CALIFORNIA PRESS
publishes bold, progressive books and journals
on topics in the arts, humanities, social sciences,
and natural sciences—with a focus on social
justice issues—that inspire thought and action
among readers worldwide.

The UC PRESS FOUNDATION
raises funds to uphold the press's vital role
as an independent, nonprofit publisher, and
receives philanthropic support from a wide
range of individuals and institutions—and from
committed readers like you. To learn more, visit
ucpress.edu/supportus.